REPORT 111

D1353439

Structural renovation
of traditional buildings

CIRIA

CONSTRUCTION INDUSTRY RESEARCH AND INFORMATION ASSOCIATION
6 Storey's Gate, Westminster, London SW1P 3AU
Tel 071-222 8891   Fax 071-222 1708

SUMMARY

Structural renovation of traditional buildings

Construction Industry Research and Information Association

Report 111, 1986 (reprinted 1994 with update amendments)

The Report examines the structural aspects to be considered in renovating traditional buildings of the 18th, 19th and early 20th centuries. It deals with the broad range of housing, commercial and industrial buildings, but it excludes monumental buildings. The approach to the structural inspection and appraisal of old buildings is presented. Detailed sections describe the assessment, repair, and upgrading of structures built of masonry, timber, cast and wrought iron, and mild steel. The design and construction aspects of foundation strengthening and retention of the façades of buildings are also covered. Sources of additional information are given.

Keywords (from *Construction Industry Thesaurus*)

Restoring; structure; traditional

Reader Interest

Architects; Builders; Quantity Surveyors; Resident Engineers; Site Engineers; Temporary Works Designers

ISBN: 0 86017 257 0   **ISSN: 0305–408X**

Foreword

To successfully retain buildings which we consider to be important for social or architectural reasons, it is essential that these buildings should also serve a useful purpose in the modern world. The preserved buildings frequently require changes in their internal planning, extensive improvements to their environmental services, possible increases in their load capacity, and often a significant extension of their life expectancy. Essential to each of these are questions of strength, stability and durability of the structure.

This Report has been prepared in the absence of comprehensive published information on the renovation of the structure and fabric of traditional buildings, principally of masonry, timber, and iron, which comprise the largest proportion of current renovation work. Because of the experiences of the authors and the Steering Group, the Report tends to reflect the knowledge and practice on building construction in the London area, but most of the information is relevant to other parts of the country. Where examples and details are given, these are for illustrative purposes and they are not exhaustive.

The Report emphasises the need for a detailed survey and appraisal of the existing building, including the implications of the proposed modifications. Clients may feel that this is an unnecessary early expense, but the savings in potential problems during reconstruction greatly outweigh the cost of this preparatory work.

It is hoped that the Report will lead to a better understanding of the difficulties facing designers and constructors, and might encourage more discussion (including planners and developers) on the practicalities and costs of the renovation and upgrading of buildings.

The study leading to this publication was carried out by Alan Baxter & Associates under CIRIA Research Project 309, guided by a Steering Group comprising:

Mr W.A. Black	Formerly Drivers Jonas
Mr M. Bussell	Ove Arup & Partners
Mr J.S. Johnston	Sinclair Johnston Consulting Engineers
Mr D.M. Lambert	Cyril Blumfield & Partners
Mr R. Powell	Formerly Trollope & Colls Ltd
Dr R.M. Lawson	Steel Construction Institute, Ascot (formerly CIRIA)

Mr R. Bowles, Alan Baxter & Associates, prepared the various sections on materials with contributions from Mr A. Connisbee and Mr J.S. Johnston. Mr F.H. Hughes, Cementation Ground Engineering Ltd, provided some of the information used in Section 6, Foundation investigation and strengthening, and Mr J. Brailsford some of that used in Section 7, Retention of existing façades. The project was funded by the Department of the Environment and CIRIA. The Report was prepared for publication by Dr R.M. Lawson and Mr B.G. Richardson, and the drawings made by Mr I.D. Clark, the 1994 update was managed by Dr B.W. Staynes, Research Manager at CIRIA.

Photographic material was supplied by the following organisations:

Alan Baxter Associates
Brailsford Engineering Associates Ltd
Cementation Ground Engineering Ltd
Curtins
Fondedile Ltd
Sir Robert McAlpine and Sons Ltd
NCL Consultants
Pinford South Ltd
S.B. Tietz and Partners

Contents

List of illustrations

List of tables

Summary

The Report examines the structural aspects to be considered in renovating traditional buildings of the late 18th, 19th and early 20th centuries. It deals with the broad range of housing, commercial and industrial buildings, but it excludes monumental buildings. The approach to the structural inspection and appraisal of old buildings is presented. Detailed sections describe the assessment, repair, and upgrading of structures built of masonry, timber, cast and wrought iron, and mild steel. The design and construction aspects of foundation strengthening and retention of the façades of buildings are also covered. Sources of additional information are given.

Introduction

At the present time in Britain, the demand for new buildings is declining, and more effort is being put into maintaining and adapting the stock of existing buildings. This is partly for economic reasons, but there is also strong social, architectural and legislative pressure to keep traditional buildings, rather than replace them with modern buildings. It is estimated that renovation, repair and maintenance currently account for 40 % of the output of the construction industry. A significant proportion of this is directly related to structural renovation.

Many professionals engaged in the building industry are experienced in the use of modern materials, choice of structural form, and in 'calculations' of member strength. However, renovation demands an understanding of the performance of buildings and materials based more on experience rather than analysis, and a willingness to regard the building itself as the primary information source. Earlier alterations and decay may have affected the character of the building, and the key to successful re-use is often the quality of the original construction.

This Report deals with building structures and materials not covered by modern codes and methods of analysis, or by text books. In the main, these are 'traditional' buildings (prior to the introduction of structural frames), which were built during the main expansion of Britain's cities in the 18th, 19th and early 20th centuries.

The Report is complementary to 'Appraisal of Existing Structures' published by the Institution of Structural Engineers [1]. It is divided into sections on traditional materials (masonry, timber and iron) and later sections on foundations and façade retention. The emphasis is on building elements rather than building forms. Nevertheless, the means of providing stability and integrity, as well as strength, are crucial in all building renovation work. To this extent, the interaction between the elements and the plan form of the building is important.

The importance of the assessment of the structure, as it is, and how it is likely to be affected by the proposed modification, is considered in Section 2.4. Uncertainty is a feature of all renovation work, in that unexpected details, deterioration, or alterations may be encountered, perhaps not until the building fabric is fully uncovered. Allowance for these should be made in the cost-estimate and programming of the work.

The Report concentrates on the structural aspects of detached and terraced housing, retail and commercial premises, warehouses and other industrial buildings typical of those found in Britain's towns and cities. Because of the variety of materials, and of the ways they were used in different forms of building and in different parts of the country, the Report covers only the more typical examples which may be encountered. The Report has been written primarily for structural engineers, architects, surveyors, and builders who may be relatively inexperienced in renovation work.

1. Historical aspects of construction

The period between 1774 and 1920 was chosen for the scope of the Report, because 1774 coincides with the introduction of comprehensive building regulations in London [2] and the start of the use of iron in civil engineering structures, while 1920 represents the beginning of 'framed' construction in steel or concrete, designed by modern calculation methods. Table 1 shows the periods during which various materials and structural techniques were in common use.

1.1 BUILDING QUALITY AND DEFECTS

Speculative housing built during this era can vary significantly in quality. Poor quality joinery in a 19th century house is likely to indicate low quality elsewhere, such as undersized flooring and lack of care in brick bonding. Stucco was often used to hide a poor quality building shell.

Industrial and commercial buildings often follow a different pattern, because more pride was taken in the works, and they were carried out by larger and more experienced builders. The tremendous improvement in civil engineering construction in the middle of the 19th century spread into building construction, and, towards the latter part of the century, all elements from the foundations upwards were likely to be better thought out and executed.

Few buildings remain in the state in which they were originally built. The quality of alterations can be poor, and a well conceived and competently built structure may have suffered later from the actions of builders or others who disguised the alterations they had made.

Many of the defects in one building can be observed in its contemporary and similar neighbours. Early builders were often cautious by current standards in their assessment of the load capacity of masonry or cast or wrought iron, but they were either highly optimistic or less concerned about the bearing capacity of their foundations. Settlement therefore often took place after construction. Many building defects respond best to a 'helping hand' rather than brutal surgery, which can generate further problems. For example, a frequent problem in domestic type structures is a lack of 'togetherness', where floors and roofs are parting company from leaning walls. Addition of some suitable ties can give the structure the restraint it needs, and allows it to respond to movements without further distress.

Defects in an existing building are often better understood if they can be monitored over a period of time. Examples may be out-of-plumb walls or cracking. It is important to understand how any defects arose, whether they have reached an acceptable state of equilibrium, and how they would be affected by the proposed work.

Most defective old buildings give plenty of warning signs before they become dangerous. In this respect, many buildings which may appear to be unsafe, according to a simplified calculation of the strengths of the individual elements, continue to resist load without apparent distress. This may be because of load redistribution, membrane or arch action, or other factors not normally considered in structural analysis.

An assessment of the existing structure therefore calls for a great deal of 'engineering judgement', and the various stages in the assessment process are discussed in Section 2.2.

1.2 BUILDING FORM

Buildings of the Georgian, Victorian, and Edwardian eras do not generally have a discrete structure, such as is found in modern framed buildings with separate cladding. As an illustration of the structural form and variety of elements on traditional buildings, Figures 1 and 2 show detailed views of a three-storey terraced house and a large industrial building, typical of the Victorian era. The cut-away sections reveal the various elements which are described in the Glossary.

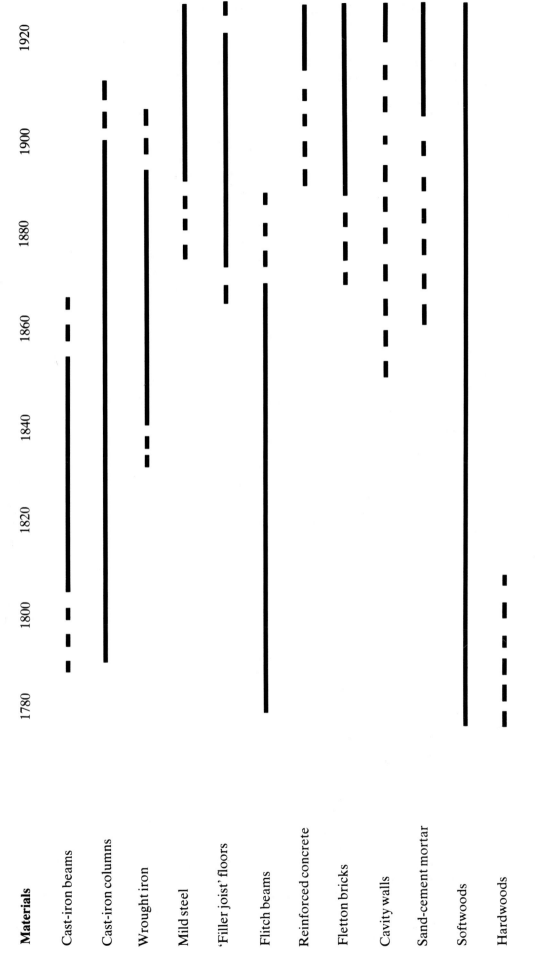

Table 1 *Historical timescale of use of different materials (solid bars denote common usage)*

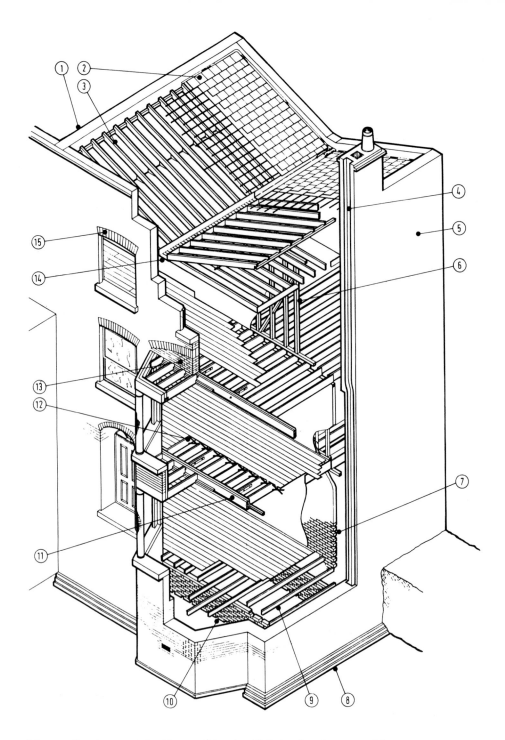

Figure 1 *Isometric view of typical Victorian terraced house*

Key
1. Party wall (neighbouring house beyond)
2. Slates or tiles on battens
3. Timber rafters
4. Flue in chimney
5. Brick flank or gable wall
6. Braced timber stud wall (as load-bearing spine wall)
7. Brick spine wall
8. Corbelled brick footing
9. Timber floor joists and boards
10. Dwarf or sleeper wall supporting joists
11. Trimmer joists
12. Herringbone strutting
13. Relieving arch
14. Valley beam
15. 'Flat' arch (normally with timber lintel behind)

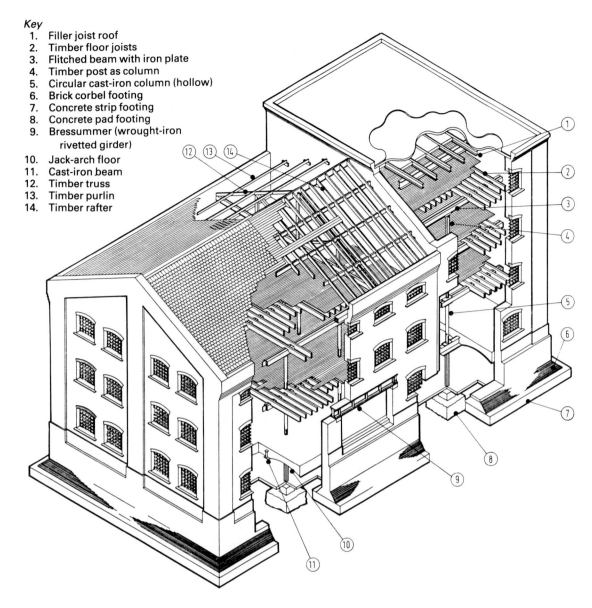

Key
1. Filler joist roof
2. Timber floor joists
3. Flitched beam with iron plate
4. Timber post as column
5. Circular cast-iron column (hollow)
6. Brick corbel footing
7. Concrete strip footing
8. Concrete pad footing
9. Bressummer (wrought-iron rivetted girder)
10. Jack-arch floor
11. Cast-iron beam
12. Timber truss
13. Timber purlin
14. Timber rafter

Figure 2 *Isometric view of typical Victorian industrial building*

These Figures show that the walls form a major load-bearing and stabilising function, but it was not until 1948 that the first Code of Practice, Structural recommendations for loadbearing walls, was published [3]. The majority of buildings therefore were not designed in the modern sense, but they were constructed with different degrees of understanding of traditional rules of thumb, leading to different structural features and form in apparently similar buildings. Terraces were frequently built in phases by different speculative builders. An architect often designed only the principal elevations, leaving the details to the builder and master tradesmen. This variety is often reflected in the floor framing system and the layout of the partitions. In some terraced houses, relatively long floor timbers may span over internal spine walls with loads over large windows or bays on the front façade carried by beams (bressummers) or brick arches. In other examples, floor joists can span between cross or gable walls. In Scotland, a large number of tall tenement blocks were built during the 19th century with relatively light timber floors supported on stone façade and internal spine walls[4].

A common feature of all of these buildings is the numerous flues in the cross and gable walls which may have caused severe deterioration of the masonry. Other problems of stability arise when these walls are not properly tied into floors, and where buildings in a terrace depend on their neighbours for lateral support. This is important where conversions to form shop-fronts have been made in an *ad-hoc* manner.

Warehouse and mill buildings often display a greater variety of load-bearing floors, because the storage and machine loads they were required to resist could have been of the order of 10 kN/m². Figure 2 shows rather more types than may be expected in one building. Jack-arch floors (Figure 3) were often used for heavily-loaded ground floors over basements. In the South, heavy timber beams on cast-iron columns were more common for the upper floors, but in the North, many all-iron beam and column systems were used.

The large warehouse buildings constructed around 1840 in London and Liverpool are fine examples of the use of masonry and cast iron. Albert Dock in Liverpool by Jesse Hartley is currently undergoing renovation, and it is the largest complex of Grade 1 listed buildings in Britain (Figure 4). The now renovated St Katherine's Dock in London was constructed slightly earlier in 1828.

There are also many other geographical variations in methods of construction and materials used. The construction of the façades in brickwork and stone are covered in detail in Section 3. The Report does not catalogue all forms of building in the period under consideration, but it attempts to present those factors which may have influenced the building's performance over its life, and how it may be affected by any proposed alteration or upgrading.

Figure 3
Jack-arch floor supported on cast-iron beams and column

Figure 4
Albert Dock redevelopment

2. Approach to building renovation

The approach to renovation of old buildings differs in many ways from new construction. The obvious difference is that the building exists, rather than being a set of drawings, specifications, and calculations. Old buildings have usually reached an equilibrium with their environment, and over their lives have distorted to accommodate the forces and other actions upon them.

Now-obsolete materials (e.g. cast and wrought iron, and lime mortar in brickwork) were often used, and current materials (e.g. timber) were used in ways which would not be likely in new design. Calculations based on inserting historical data on materials strength into modern design formulae may give the impression that the designer has fully appraised the structural adequacy of the building. On the one hand, it may not be possible to justify the structure by calculation despite the building demonstrating the contrary. On the other hand, detailed calculations may ignore other important factors which may seriously weaken the structure (e.g. loss of bearing or tie action).

There is no substitute for a proper survey, inspection, and appraisal of the building as it is, to establish how it is likely to be affected by the proposed modification. The stages in the appraisal process are described in Sections 2.2 to 2.4.

2.1 DIFFERENCES BETWEEN OLD AND NEW CONSTRUCTION

The key differences between old and new buildings which should be recognised are:
1. Absence of a discrete structure: the elements of the building combine to prevent collapse in its existing state, despite the lack of a discrete and calculable structure. This is particularly true of vertical elements where the load path to the ground may not be direct.
2. Quality of the original construction: this is something which exists, and which cannot be specified and subjected to control. The results of an assessment of quality can have a significant effect on the approach to further alterations.
3. Previous modifications: many old buildings have been modified a number of times, often on a piecemeal basis, so that their combined effect can make further work more difficult. An appreciation of the form and scale of these modifications is essential.
4. Change and decay: all buildings are subject to gradual decay from deterioration of the materials used, weathering, or lack of preventative maintenance. The scale of the decay influences the scale of the renovation work, and it determines how the life of the completed building may be extended.
5. Existing use: it is not unusual to find the building occupied at the time of the initial survey and assessment, which can hinder any investigation. Proposals should also consider the rights of the adjacent owners, particularly with regard to the stability and weather-tightness of party walls.

2.2 SURVEY OF THE EXISTING BUILDING

The term 'structural survey' is often used in surveyors' parlance to denote the survey of the whole building (e.g. for mortgage purposes), and it should not be confused with the 'survey of the structure', which is the engineering appraisal of the load-bearing elements of the building. The term 'survey' is taken here to mean those aspects within the province of the structural engineer.

At the outset of a renovation project, and ahead and independently of the rest of the design team, the engineer should start to collect information about the building, its form, its materials, and its history. A visit to the library to search out old maps and historical records is a worthwhile starting point. This may well show the former terrain and old watercourses. The Code of Practice for site investigation [5] lists many of the sources of information.

Equipment such as a sharp probe, a hammer, a cold chisel, a nail bar, a torch, a plumb bob, a moisture measuring device, a mirror, a spirit level, long and short tapes, and a folding ladder are essential tools for the physical survey.

After the survey and initial assessment, it may also be necessary to employ specialists to determine the properties and constituents of the materials used, and to advise on any remedial or protective measures. Further information on the inspection of the foundation conditions is given in Section 6.1.

The stages in the survey are:

2.2.1 Collect general information
Date of construction and of adjacent buildings.
Boundaries of, and access to, site and existing building.
Permission from local authority and owners.
Planning requirements – conservation area, environmental factors.
Existing records – local authority, owner, maps, etc.
Present and previous uses of building.
Lease agreements – freeholders and leaseholders.
Adjacent buildings and owners – party wall agreement.
Vegetation, drainage, and risk of flooding.

2.2.2 Survey the structure of the building
Record plans of each floor and elevations of walls.
Record the form of the construction and principal load paths.
Identify the structural materials used.
Record the dimensions of the structural sections or elements.
Note signs of distress and whether they are continuing or stable.
Expose and record typical connections between elements.
Investigate the stabilising elements of the structure (e.g. ties, crosswalls).
Note the form and scale of previous modifications (see below).
Inspect foundations (see Section 6).
Assess the *in-situ* strength of the materials used (see relevant materials section).
Note the location of stairs, chimneys and flues.
Investigate 'dead areas' revealed on floor plans (e.g. blocked up ducts).

2.2.3 Record defects, deterioration and alterations
Note bulging, settlement and other distortion of walls.
Record the position of cracks, and those which appear to be 'widening'.
Note the 'feel' and sagging of floors.
Inspect roof voids, wall plates and joists for possible deterioration.
Extra floors or walls added or walls removed?
Fire places and chimney breasts removed or altered?
Windows and doors as original or altered?
Note other extensions and alterations to the building.

2.3   ASSESSMENT OF THE EXISTING STRUCTURE

Section 2.2 dealt with the documentary and physical parts of the structural survey, and it represents the information on which future decisions depend. A skimped survey may lead to problems when the contractor is on site. However, unexpected construction details may still be encountered at the construction stage, and they should be allowed for in the contingency sum.

Sufficient drawings should be made to decide how the structure behaves. Sectional elevations and isometric sketches are often more useful than detailed plan drawings at this stage.

In general, the initial structural assessment should supply answers to the following questions in order of importance:

1. Is the structure generally robust and stable? If it is not of reasonable quality, re-use may not be feasible without major alterations or repairs.
2. Are there signs of continuing distress? If so, their causes must be diagnosed and remedial measures considered.
3. Are there signs of past distress?
4. Is there a likelihood of significant defects remaining undiscovered until the construction stage?

At this stage, the designer is probably aware of the proposed use of the building, and this should be taken into account in the assessment. The loading for which the original structure was designed, or to which it might reasonably have been subjected during its life, is also useful information (see Section 2.4). From a visualisation of the load paths through the building, the probable stress levels in the various elements may be estimated. Considerable judgement is required in interpreting these values before deciding what is sound or potentially dangerous, taking account of hidden or possible further deterioration. In many cases, the stiffness of the structure (as distinct from its strength) is the governing factor.

Some common faults which may be encountered are:

1. Walls
   thick stone or brick walls may be rubble filled
   lintels may be non-existent and loads applied to door or window frames
   external and internal brick skins or intersecting walls not bonded
   large flues in brickwork which has deteriorated
   chimneys breasts removed in lower storeys
   bonding timbers present and subject to rot and shrinkage
   openings through stud partitions have destroyed framing action
   out-of-plumb walls concealed by finishes.

2. Floors
   sagging concealed by finishes
   no tying into walls
   deteriorated wall plates
   cut-out sections with inadequate trimming.

Cracks in the structure require careful consideration, and their pattern and widths should be recorded. The evidence of continued movement may be determined by installing 'tell tales' across suspect cracks. It is advisable to take readings over as long a period as possible, including records of temperature and dampness effects. The diagnosis of the cracking pattern can lead to identification of the source of the movement (e.g. settlement, shrinkage, or relative deflections of floors). The settlement of the foundations of old buildings often resulted in cracking and distortion over a long period. Therefore provided the loads on the foundation do not increase, and the stiffness of the wall does not reduce, further movement may not be significant (see Section 6).

2.4 ASSESSMENT OF THE STRUCTURE FOR ITS NEW USE

Structural adequacy broadly means the structure is stable and possesses sufficient strength and stiffness for its intended purpose. Having established the condition of the existing structure, it is necessary to consider the implication of the proposed modifications and change of use.

Meeting new load requirements frequently calls for an increased load capacity, and the existing structure should be assessed for its ability to support this. Adjustment of load factors to justify lower factors of safety in existing buildings may be appropriate for relatively modern concrete and steel frames, but any such approach for more traditional building and materials should be related to the likelihood and form of any defects and their influence on the strength of critical elements (see Section 2.5).

No statutory guidance on load requirements was published before the 1909 London Building Acts[6,7], and the text books of the day were often used. The live loads specified in the 1909 Act were relatively high compared to those revised subsequently in the 1930 Act[7] and those used in the design of modern buildings [8]. However, many builders in the 19th Century (and even subsequent to these Acts) did not adopt such onerous loads for the majority of domestic and commercial buildings, which were sized by rule of thumb. For this reason, some elements may be adequate, but others may be overloaded, and an assessment of their strength and stiffness is important. It is therefore unwise to proceed on the assumption that a building has the live load capacities given by the 1909 Act.

If the building is found to be inadequate for its new use, the feasibility of the scheme and its various options should be reconsidered. It may be necessary to strengthen or replace parts of the existing structure, or to provide a separate structure. Particular attention should be paid to the load transfer to critical elements resulting from openings or other modifications. In general, any additions or changes should not disturb the existing load path.

Removal of floors or stabilising walls, even temporarily, may jeopardise the ability of the structure to withstand wind or other lateral loads. Conversely, the load-carrying capacity of walls can be greatly improved by providing better tying action between the walls and floors. It is important for the designer to consider the feasibility of the temporary structural works needed to maintain stability during the building operation and to liaise with the contractor on where the work has to be performed in a certain sequence.

Other factors which influence the building in its renovated form are the need for fire resistance, acoustic and thermal insulation, and division into compartments. In general, these improvements are met by applied finishes or partitions which, in turn, may increase the weight on the structure.

In many cases, the functional requirements of the proposed building cannot be met economically, and complete rebuilding, perhaps retaining the façade for aesthetic or planning reasons, is the only solution. Façade retention is considered in detail in Section 7.

2.5 MEANS OF SAMPLING AND TESTING

After the preliminary assessment, it is often desirable to obtain more information about the materials used. This may extend to a determination of their mechanical properties (e.g. by metallurgical and strength tests on metals, stress grading of timber, and analysis of mortars). Whether or not detailed sampling is practical and economic depends on the scale of the structure, its form, its quality, and the potential benefits.

Quality control in the 19th century was much poorer than it is today, and the scatter of strength test results will be considerable. Interpretation of the 'design' stresses which may be used is subjective. In general, the permissible stresses used in the 1909 London Buildings Acts or in other contemporary text books suffice. Any proposed increase in these stresses should be carefully considered and discussed with the building control officer.

Examples of the methods of test which may be used are given in the relevant materials sections of this Report and Reference 1. To obtain realistic results, a reasonably large number of test samples should be taken to include both geometric and material variations in the structure and any likely defects. The connections can usually only be assessed visually. The thickness of steel plates or welds may be measured by ultrasonic devices, but this is rarely successful for cast iron and wrought iron. Many materials tests can only be carried out by specialists. These tests may therefore be expensive. It may be advisable to take additional *in-situ* measurements to confirm these results.

Load tests on representative areas or bays may also be carried out in extreme cases to justify the capacity of the floor. Tests may be to service load to check deflection, load distribution, etc., or to overload to assess its safety[1]. Small loaded areas may be heavily restrained by neighbouring sections, and this should be taken into account in the interpretation of the load test.

Although 'spare' capacity may exist in certain components, the strength of all the elements, and in the foundations and connections in particular, is rarely consistent. Testing therefore tends to be 'confirmatory' rather than 'indicative'.

3. Masonry construction

Masonry construction in both brick and stone has developed as a traditional craft, and its various forms are well presented in References 9 and 10. This section deals with masonry in terms of the materials used and its common features, leading to an assessment of its structural performance.

3.1   DIFFERENT TYPES OF BRICKS

Many different kinds of brick were in use during the 18th and 19th centuries, ranging from strong and durable 'engineering' bricks produced in large kilns to 'common' bricks produced locally, or even fired on site in 'clamps' (see Glossary). It is not the purpose of this Report to compile a complete survey of brick types and regional variations. Here is a brief explanation of the general names and descriptive terms which most frequently arise.

COMMON BRICKS were made from locally occurring or readily available brick clays, little attention being devoted to evenness of firing or regularity of form. Such bricks were usually unsuitable for facings, but were strong and hard enough for all common purposes. Today, the name usually refers to plain Fletton bricks which, although manufactured with precision of form and composition, are used almost exclusively in work concealed by decorative finishes.

STOCK BRICKS derived their name from the use of a stock board in moulding bricks by hand, but the term 'stock' now has no accurate definitions other than that it excludes bricks without 'frogs'. It is often loosely applied to common bricks in general or to the characteristic yellow-brown 'London stocks' produced on both sides of the Thames. Well-burnt stock bricks suitable for use in external walls are hard, durable, and have a characteristic ring when one is knocked against another.

PLACE BRICKS are those which, having been outermost or furthest from the fire in the clamp or kiln, have not received sufficient heat to burn them thoroughly. Consequently, they are soft, uneven in texture, and of a red colour. At best they are suitable only for non load-bearing partitioning. The poorest examples are sometimes more akin to baked clay, and they readily deteriorate on being exposed to the weather.

BURRS AND CLINKERS are over-burnt bricks which have partially vitrified, distorted and run together in the clamp or kiln. They are hard, durable, and suitable for outside work in rough walling.

RUBBERS OR CUTTERS are lightly burnt to be soft enough to cut or rub down very evenly and uniformly. They have no frog, and they are usually oversize to allow for rubbing. Consequently, they are heavier to handle.

ENGINEERING BRICKS are strong, dense, and smooth-faced. They are obtained by burning bricks to vitrification. Staffordshire Blues are the best known, though even stronger red varieties were made at Accrington, Bristol, and in parts of Yorkshire etc. Their use developed from civil engineering applications in the mid 19th century.

GLAZED BRICKS have a glazed surface, and they are usually white. They are commonly used in decorative, low maintenance façades, or internal lightwells. In both cases, they may have been added later to a brick facing.

GAULT BRICKS are hard, close-textured pale buff bricks manufactured from the chalky clay of the same name in Cambridge, Bedford, Essex, and parts of Kent. Gaults were among the first bricks to be mass produced by machine, and wire-cut perforated varieties date from the 1850s.

FLETTONS, named after the village near Peterborough, were also early machine-made bricks moulded from the Oxford clay of the Bedford, Buckingham, and Northampton districts. Their normal colour is pink – usually patched where bricks have been stacked on top of one another in the kiln. They are hard, regular-shaped bricks of great use in load-bearing, non-facing construction.

SAND-LIME BRICKS are formed from chalk lime and 5 to 10 % sand, ground together, then moistened and machine pressed into bricks. These bricks are cured by exposure to high-pressure steam for about 10 h. Under these conditions, the sand and lime combine chemically to form strong, exceedingly regular-shaped bricks. These are more commonly found in more modern buildings, and they tend to shrink gradually, whereas clay bricks, hot from the kiln, are prone to 'early-life' expansion.

CONCRETE BRICKS are cast in moulds. They differ from other forms of brick in that the aggregate is physically bound by cement rather than chemically combined with the other constituents. Concrete bricks have similar characteristics to sand-lime bricks, and were only produced in quantity after 1920.

The bricks in buildings of the 18th to early 20th centuries would be an assortment of qualities. Sound and regularly-shaped bricks were the more expensive, and they were used on the weathering faces of buildings, though the inner, concealed body of a wall commonly comprised the softer and cheaper place bricks. Rubbers were frequently used for arches and decorative courses.

The mixture of bricks employed in external walls is readily visible in London, where the well-burnt stock brick takes on a characteristic yellow grey colour, but the place brick remains pink. On stripping internal plaster off external walls of houses, it is invariably found that the facing bricks are backed internally by pink place bricks. Party walls often comprise place bricks throughout, as do areas of external walls which were decorated with stucco.

3.2   TYPES OF MORTAR

The principal reasons for laying bricks or stones in mortar are that bedding in mortar encourages a uniform transfer of load through elements whose irregularity or distorted form might break, even under quite light pressures. Irregularities can thus be absorbed without variation in the coursing. The mortar also acts as a gap filler to keep out the weather.

Friction and adhesion between mortar and brick or stone prevents dislocation of the individual elements, and it enhances the load capacity of the wall. However, the tensile strength of the mortar is small, so the wall effectively behaves as a gravity structure.

Because all 18th and 19th century mortars are limestone or chalk based, only the common lime mortars and Portland Cement are discussed here in detail. Pozzolanic mortars were sometimes used.

3.2.1 Lime-based mortars

The understanding of slaked, or burnt lime, possessed by the Romans, disappeared with the decline of the Roman Empire. Lime manufacture became pretty much a matter of trial and error until the 1750s, when John Smeaton formalised research into mortars for use in his reconstruction of the Eddystone Lighthouse.

By heating limestone (calcium carbonate) in a kiln to around 850 C, carbon dioxide and water are driven off to leave quicklime (calcium oxide).

PURE OR FAT LIMES, such as may be manufactured from chalk, are readily slaked with water, with the evolution of considerable heat and a doubling of volume. These limes set or harden entirely by absorbing carbon dioxide from the atmosphere to form a soft crystalline carbonate of lime in a process known as carbonation. The slow gain in strength makes these mortars unsuitable for rapid construction.

Limestones containing clay impurities can be burnt to form calcium silicates and aluminates. These slake with less violence than fat limes, and they show less volume increase. On addition of water, these so-called hydraulic limes set and harden chiefly by independent internal crystallisation. Any free lime present hardens by absorbing carbon dioxide. Generally speaking, the greater the clay impurity, the faster the setting time and the less carbon dioxide is required. Hydraulic limes are generally stronger than fat limes, and they were used for structural purposes.

Both plastering and structural limes were traditionally mixed with sand (around 1:3 by volume) to increase the bulk of the material, to assist setting, and to reduce drying shrinkage. Any aggregates used in mortar should be free of salts and other impurities, the presence of which might retard the setting of the lime or otherwise weaken the mortar.

Although several patent products were marketed during the 19th Century as 'cements' (Parker's 'Roman' cement, Frost's 'British' cement), they were really only varieties of hydraulic limes containing differing proportions of clay.

3.2.2 Portland cement

The manufacture of Portland cement in the form we recognise today dates from 1824, and its history is reviewed in Reference 11. The basic ingredients were still limestone (originally chalk) and clay, but in the proportion 2:1. They are ground finely together and burnt between 1200 and 1400 C to incipient vitrification, resulting in the formation of clinkers. These clinkers are then finely ground to form the finished cement. Addition of water triggers off a series of chemical reactions which result in the crystalline formation of hydrated tri-calcium silicates and aluminates.

Until approximately 1900, masonry in buildings was mainly constructed with lime mortars which thereafter slowly gave way to cement mortar. By the 1930s, cement mortars were most prevalent. They invariably contained a proportion of lime to improve workability during construction.

Cement mortars are comparatively hard and brittle, and they have higher crushing strengths than lime mortars. As a consequence, walls built of cement mortars have a greater tendency to crack when subjected to distortion – either by brittle fracture of the mortar or, if the brick is the weaker material, by the snapping of bricks. Relatively small distortions can cause visible cracking. By contrast, lime mortars are more plastic, and they can accommodate long-term movements.

3.2.3 Other mortars

Prior to the time when Portland cement was cheaply and readily available, other materials exhibiting cementitious properties were used to achieve strengths greater than those of traditional lime mortar. One of the principal materials was the ash from coal-burning operations in industrial processes. This flyash was highly variable, and, because of its sulphur content, it was acidic, in contrast to lime. Metallic elements tend to corrode in flyash mortar, which is characterised by its dark appearance. It was used throughout the 19th century and up until the 1930s in areas such as South Wales, where the corrosion of wall ties is a notorious problem.

Naturally-occurring pozzolanic materials were sometimes used. Modern pozzolans produced from coal ash are much more refined and less variable.

3.2.4 General problems of mortar

Although text book theory indicates that lime-based mortars can be hard and durable, it was common practice for builders to skimp on the lime content of a mortar and to use unclean, poorly-graded aggregates. Housing in London in the 19th century (and particularly during the 1870s) was notorious for the prevalence of poor mortars, and there were articles in the national press about this state of affairs, following a number of building collapses.

The behaviour of loosely bedded stacks of bricks in load-bearing situations is discussed in Section 3.7.3. The principal purpose of mortar in the 18th and 19th centuries was to provide a stable material into which irregularly shaped building units could be bedded one on top of another. It also provided a joint filler. For this latter purpose, lime mortars were found to be adequate, because an external weathering surface (pointing) was provided.

One of the main beneficial features of lime mortars is that, being generally plastic rather than brittle, they are able to accept quite large deformations and distortions without the opening of bed joints or the splitting of bricks. This probably explains why any expansion of the bricks early in their life caused little visible cracking compared with modern masonry.

With regard to chemical stability, the most common aggressor has probably been soluble sulphates – either from groundwater or from the bricks. These slowly attack ordinary cement under continuously wet conditions, and they give rise to expansion and softening of the mortar. In severe cases, total disintegration of the mortar can occur. Render, stucco, and other cementitious finishes can be affected by sulphate attack.

Manufacture of building limes is carried out at lower temperatures than for cements. This results in a different chemical structure from basically the same ingredients. The sulphate-susceptible compounds which are relatively abundant in cement are scarce in limes, so that sulphate attack is less of a problem in lime mortars.

The bulging of exposed walls containing lime mortar is frequently put down to sulphate attack, but in fact it seems more likely to be associated with a number of other processes such as the absorption of carbon dioxide into the mortar during setting, the premature loading of fresh brickwork, the non-uniform construction of a wall section, or lack of lateral restraint.

3.3   BRICKWORK AND ITS COMMON DEFECTS

3.3.1 Brickbonds
The art of bricklaying had been well understood by the 18th century, and the qualities of different bonds were known (Figure 5). By far the most popular bonds were the 'English' and 'Flemish' bonds, of which the latter became the most widely used in 18th and 19th century house construction. In Flemish bond, each course comprised alternate header and stretchers overlapping with alternate headers and stretchers in the course above and below. The headers were intended as full bricks to bond the facing work into the backing work, the latter usually being constructed of common bricks. It was considered to be more decorative than the English bond [10].

Also popular was a bond known as 'Flemish garden wall' in which headers were separated by three stretchers. This represented a substantial decrease in the number of headers and a lesser bonding of the facing and common brickwork.

English bond consists of alternate courses of headers and stretchers. It is considered to be very strong, because of the absence of straight joints within the wall, but it is more difficult to lay, and it is more expensive than other bonds. 'English garden wall' bond uses either three or five courses of stretchers to one course of headers, and it is one of the most popular of all brickwork bonds, particularly in the North and Midlands. Many other brick bonds were used with regional variations in their names.

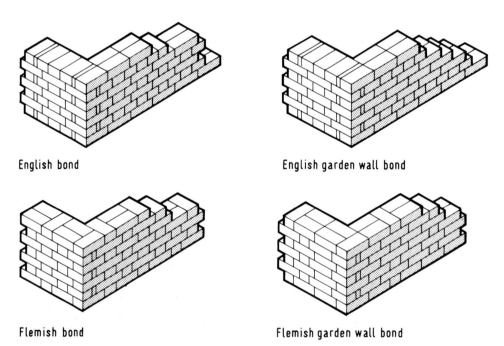

English bond   English garden wall bond

Flemish bond   Flemish garden wall bond

Figure 5   *Illustrations of brick bonding*

3.3.2 Wall slenderness and restraints

Regulations [2] for the thickness of external and internal or party walls depended on the number of storeys and the storey height. The minimum thicknesses specified in the Metropolitan Buildings Acts of 1844 for 'first class' buildings are reproduced in Figure 6. The slendernesses of the walls were cautious by modern design standards if the wall was restrained effectively at a storey height of 8 to 10 ft (2.5 to 3.0 m) in a typical house.

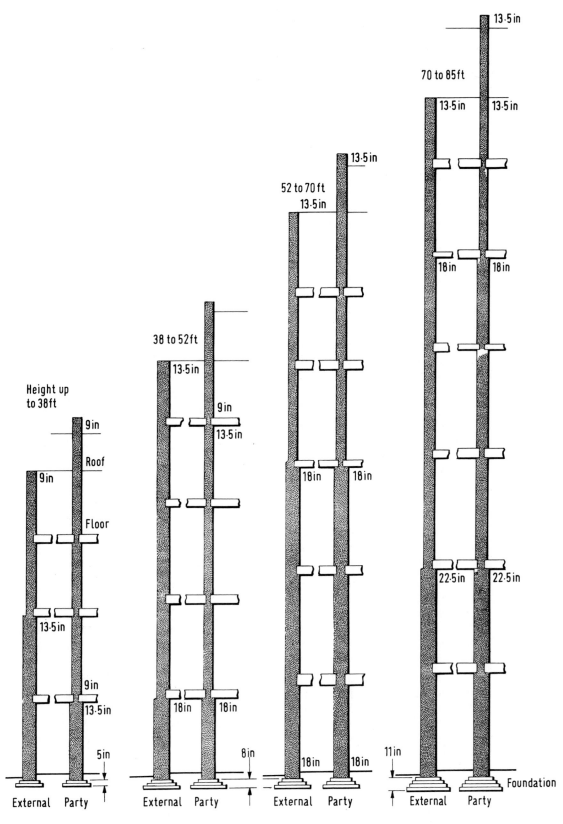

Figure 6 *Thickness of walls in accordance with the London Building Acts of 1844*

However, the restraints at floors and crosswalls were often weak or non-existent, and the walls were often constructed of poor materials, or they were not as thick as required. Consequently, instability of these walls, particularly at gable ends or adjacent to large openings is an important consideration. The common defects in brickwork are discussed in the following sections, and its appraisal is covered in Section 3.7.

3.3.3 Improper bonding of bricks

When front elevations were faced with gauged or rubbed bricks laid in thin 'putty' joints, the depth of individual facing courses was less than in the back-up 'common' brickwork. The result was that only occasionally could the wall be bonded by the introduction of uncut headers from the facing skin into the back-up skin, when the courses chanced to fall evenly.

More common and less forgivable examples of the same weakness occur where facing-brick headers were deliberately cut in half (called 'snap headers'). Headers were provided only occasionally to give a nominal bond between facing and backing skins. It was common for the outer face brickwork to be constructed later, and the headers were often poorly bonded into pockets formed in the preceding work (see Section 3.3.4).

The inner skin of poorer bricks, often containing built-in timbers, might be carrying (and restrained by) floor loads. Any compression or contraction of this inner brickwork may lead to shearing of the few headers, even if bonded, causing delamination of the outer skin, which is then unrestrained. The problem of built-in timbers, and other features leading to distress, is covered in Section 3.6.

Rubble-filled piers, columns, or thick walls may also be encountered, the outer skin of which might have bulged under the lateral pressure of the fill. Collapse of improperly bonded brickwork may be triggered by any disturbance during rebuilding operations.

3.3.4 Improper bonding at wall intersections

Party walls were invariably plastered on both sides, and they were also largely protected from the ravages of the weather. Accordingly, builders frequently employed a poorer quality both of materials and of labour in their construction. Often, party walls were erected in advance of the more important exposed elevations, possibly by unskilled or apprentice labour, while the 'face work' was constructed with greater care for appearance as a separate operation by more experienced hands. The junction between the party and external walls was rarely bonded, and more often a straight butt joint or a rebate was formed with no attempt at bonding at all. In neither case is effective restraint given to a wall between floor levels, and vertical or toothed fractures in the corners of such junctions are commonplace.

3.3.5 Use of poor materials

Bricks of the same composition and contemporary firing can vary in quality, depending upon their position in the clamp or kiln, and mortars may contain variable proportions of lime.

Textbooks always advised on the use of sound, well-burnt bricks for all major walls, but the reality in speculative development was usually the employment of the best bricks where appearance mattered and the cheapest where work was concealed. From this arises the anomaly that the back-up brickwork, which tends to carry the most load from floors, beams, trimmers etc., was constructed from the poorest bricks. Party walls and external stucco-finished walls were commonly constructed with poorer bricks throughout.

As regards the differential behaviour under load of underburnt 'place' bricks and well-burnt facing, it seems likely that greater deformation under stress should be expected in the softer, underfired bricks. When a wall is eccentrically loaded by floors and roof, local stresses on the inner face may be considerably greater than those on the outer face. A greater consolidation of lean mortar is therefore to be expected on the inner parts of a wall, leading to bulging.

3.4 FOUNDATIONS FOR BRICKWORK

19th century publications on building construction show a theoretical appreciation of foundation 'design' to the extent that the spreading of loads was recommended at the foot of a wall by means of corbelled brickwork or wide mass concrete strips or both. The spreading of concentrated loads was achieved with inverted arches, and thoroughly bad ground conditions were tackled with piles, reinforced rafts, etc.

In buildings of the finest quality (presumably where an architect was employed), and in industrial property and large-scale public works (presumably where engineers were employed), foundations were usually reasonably well constructed. However, the architect often designed only the principal elevations, leaving the remaining work to the builder. The speculative building of the 18th and 19th Centuries was often constructed using only a nominal spread footing bearing, with the least amount of excavation to virgin soil. The rules concerning the width of foundations in relation to the thickness of the wall supported were often applied as a maximum or norm rather than a minimum requirement. The Metropolitan Building Act of 1844 [2] also allowed the tops of foundations to be only 2 in. (50 mm) below the basement floor level (see Figure 6 page 23).

3.5 STONE MASONRY

Stone masonry required more skill to build than brickwork. Bricks, being all of the same size, were laid according to a regular pattern, whereas with each stone considerable judgement was required so that it might be laid in the best way. Masonry was classed either as 'ashlar' or as 'rubble'[12].

Ashlar masonry was built with blocks of stone very carefully worked, so that the joints did not exceed 3 mm in thickness. It was the most expensive class of masonry built, and it depended for its strength upon the size of the stones, the accuracy of the dressing and the perfection of the bond, but hardly at all upon the strength of the mortar. Because the joints were so thin, the mortar should be very fine and free from grit.

Coursed ashlar consists of blocks of the same height throughout each course with regular horizontal joints. Examples of coursed ashlar with different backing materials are shown in Figure 7.

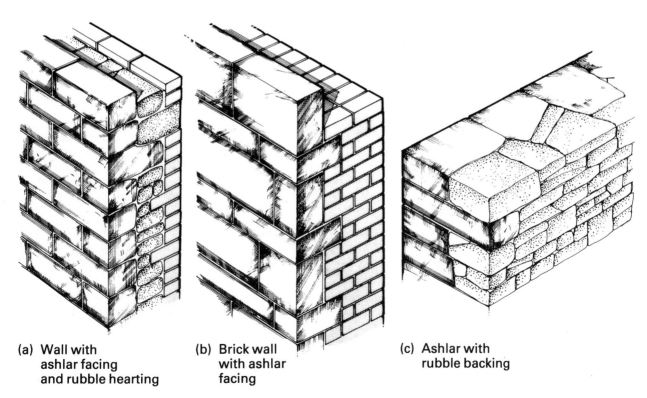

(a) Wall with ashlar facing and rubble hearting

(b) Brick wall with ashlar facing

(c) Ashlar with rubble backing

Figure 7 *Examples of stone masonry construction*

Random ashlar walls were built with rectangular blocks of all sizes and dimensions. It was a cheaper form of ashlar, because it enabled a larger proportion of the stone to be used as quarried, although it was more difficult to build with successfully.

Rubble masonry was built with roughly dressed blocks of stone. The beds and joints in rubble work were variable in thickness, and the strength of the walling depended greatly upon the mortar.

The faults and failures discussed in Section 3.3, relating to bonding within and between brick walls, can equally well occur in stone masonry construction. It was normal for masonry construction to consist of an ashlar or squared rubble facing, bonded back into material usually of poorer quality (e.g. random rubble or brickwork). In this form of construction, the coarser and more numerous joints in the brick or random rubble backing, as compared with the finer and fewer joints in the ashlar, might result in differential settlement.

Another common (but poorer) form of construction comprised a facing skin of ashlar and parallel backing skin of brickwork or rubble masonry bonded together across an intervening cavity. This cavity was often rubble filled with small loose stone chippings and broken blocks, frequently with little or no binding mortar. Delamination under these circumstances becomes even more serious as a result of the surcharge arising from this loose filling.

In later construction, some of the bonding problems were overcome by building the facing work independently to the backing but tied in with iron or steel cramps and dowels. A variety of tongues, grooves, and channels were developed to improve the bonding or interaction of the stone units. These were designed both for direct interlocking of one stone with another and for indirect connections via tightly fitting plugs or loosely fitting dowels and chases run with lead or mortar.

Rubble-filled walls, often with poor bonding between the facing and backing brick or stonework, were a feature of many, even quite tall, buildings up to the late 19th century.

In Scottish tenement buildings, the façades were usually constructed with two skins of sandstone, and the walls were 600 mm thick, reducing to 300 mm at a line of windows. The leaves were tied occasionally by through stones (or in-bands), and they were rubble filled. Gable walls were about 450 mm thick, of cellular construction, and sometimes accommodated chimney flues. They were capable of carrying little more than their dead weight.

3.6   OTHER MASONRY FEATURES

3.6.1 Hollow or cavity walls

The advantages of hollow or cavity wall construction began to be realised after the middle of the 19th century, presumably coinciding with a general awareness of the causes of dampness in walls.

Before the days of calculated masonry, cavity construction was very much an addition to the main structure of a wall [10]. The usual wall thicknesses were adopted for the inner 'leaf', the cavity being enclosed by an extra facing leaf, restrained with special ties built into the inner work. The form of wall construction was expensive to erect, and it is seldom found before the 20th century, except in the finest quality houses. Indeed, it did not reach parity with solid construction until the 1930s.

Early ties were made of cast iron, wrought iron or specially shaped clay bricks. It was good practice that the metal ties should either be galvanised or coated with hot tar, then dusted with sand to prolong their life in the wall. All the tie forms recognised the importance of shedding cavity water off the ties, so they are twisted or bent in a variety of ways.

With metal ties, differential movements between the facing and the backing brickwork can be accommodated to some extent by bending or rotation of the ties. Special bonding bricks, being both brittle and weakened by perforations, are similar to the poorly-bonded solid walls where differential movement can rupture the occasional header bricks within the wall.

This freeing of the facing skin from the backing work is the single worst defect that may be encountered with cavity construction, whether by fracture or by corrosion of the ties. Cavity walls, particularly those of older construction, should therefore be treated with some caution when assessing them for alteration or long-term retention.

3.6.2 Built-in timbers

Bonding or coursing timbers were often used behind the facing skin or in internal walls to improve the bond and to spread out loads from higher courses on slow-setting weak mortars. Sometimes timbers were built in for joinery fixings. Wall plates, set into the brickwork, were used to support the floor joists.

The natural cross-grain shrinkage or compression of bonding timbers can result in wall movements, particularly bulges in external walls. More serious is the risk of timber decay from rot or insect attack, which can cause serious structural problems, particularly in small section piers. Not only is the effective section of the brickwork reduced, but it is also subject to eccentric loading.

A typical masonry pier between sash windows in an 18th or 19th century house may have a substantial part of its area taken up by sash boxes and shutter recesses, and it may have built into it main timber floor beams as well as wall plates and bonding timbers holding poor quality masonry together.

Any decision on removal of built-in timbers in external walls should balance the seriousness of the existing problem and the risk of future problems against the major disturbances to the structure from the timber to be removed and replaced by tightly-fitting bricks. The scale of the work, as well as the nature of the existing structure, influences this difficult choice.

3.6.3 Arches

Arching is one of the oldest methods of bridging an opening. Arches are so arranged that an applied load forces one element to bear against another in pure compression as a wedge. These wedges not only have to be supported by a vertical reaction equal and opposite to the applied load, but they also have to be restrained from spreading apart by an adequate buttressing load perpendicular to the applied load. Some examples of modes of failure of common arch types are illustrated in Figure 8, together with definitions of the main elements. A useful modern reference to the analysis of arches is by Heyman [13].

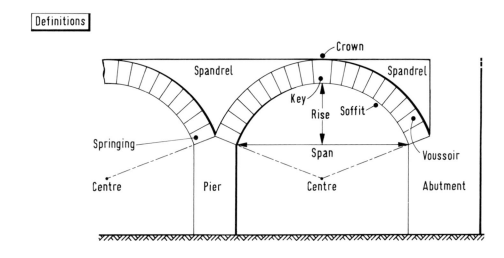

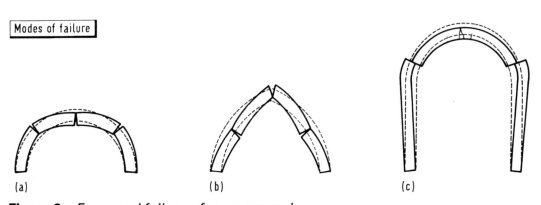

Figure 8 *Form and failure of masonry arches*

The efficiency of an arch is greatly affected by its shape. Certain parabolic forms, when uniformly loaded, ensure that there is uniform compression across any section of the arch. Some gothic arches approximate to these parabolic arch forms, and even semicircular arches are a reasonably efficient arch shape.

Such arch shapes which have large rise:span ratios remain remarkably stable, even when their abutments are displaced significantly. Displacements are accompanied by the opening of cracks as the compression relaxes, allowing hinging to take place. The arch can thus be said to have spread but not collapsed. Quite alarming displacements can take place in such arches before collapse occurs.

Generally speaking, as the rise of an arch decreases, the greater is the lateral load required to prevent spreading, and consequently the greater is the compression in the arch members. Furthermore, the arch becomes more sensitive to movement of the supporting structure.

Flat or shallow arches in brickwork are often found above window or door openings (Figure 9). The flat arch can resist loads because the line of thrust is within the arch, and the outward forces are resisted by the buttressing action of the adjacent walls.

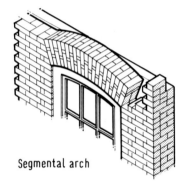 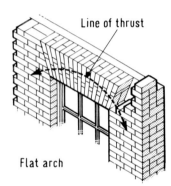

Figure 9
*Forms of arch
above windows*
Segmental arch   Flat arch

3.6.4 Domestic chimneys

Chimneys were traditionally constructed with not less than 4 in. (100 mm) of masonry separating flues from one another and from the interiors of the properties they served.

It was essential to flue design that there should be a 'gather' or reduction in section above the fireplace opening (to encourage the drawing of air upwards),also that the flue should turn smoothly to right or left (both to prevent direct ingress of rain into the fire and to allow the construction of a fireplace in the same position on the storey above). One common arrangement is where fireplaces and flues are constructed back to back against a party wall, as in Figure 10.

The gathering and turning of the flues is formed by stepping and corbelling the flue walls as necessary, the unwanted voids usually being filled with poor quality broken bricks or rubble and often with the omission of mortar bedding.

An alternative arrangement of chimneys is where the flues are contained almost entirely within the body of the adjoining wall. Party walls in buildings of five or six storeys receiving, say, two flues per storey from each of the adjoining properties, might therefore contain a total of 20 or 24 flues distributed along some 6 to 10 m of the length. Such a wall, containing so many snaking flues and rubble pockets, relies heavily on proper bonding for its continued integrity.

In both arrangements, despite textbook advice to the contrary, it was easier to build the flue structure with an external envelope laid in stretcher bond, and the 'withes' or flue dividers stack-bonded between envelope walls. Because 'pargetting' or rendering of the flues internally was mandatory, this defect is often not apparent while the pargetting remains intact, except by examination of the bond on the face of the chimney breast. Snap headers were frequently used, so that the appearance of adequate bonding is not necessarily a sign of good construction. In extreme cases, the facing and backing skins might have become detached.

Repeated heating and cooling of chimney stacks and the aggression of hot coal gases and sweeps' brushes have almost certainly loosened or removed the flue lining, and they have thus exposed the mortar joints of the brickwork beneath to attack.

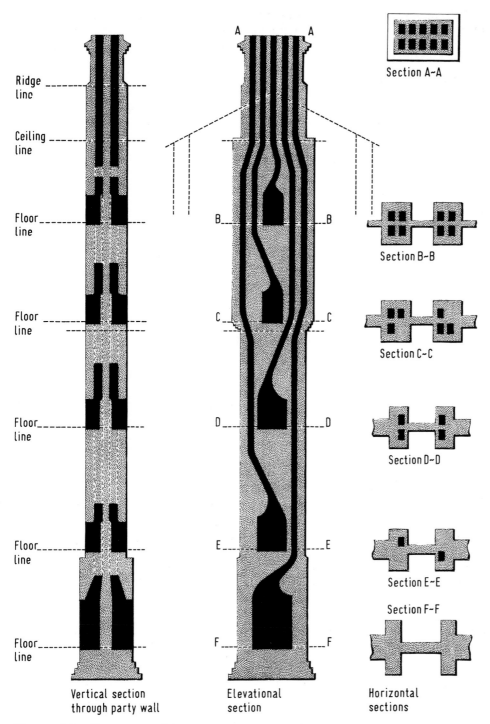

Section A~A

Section B~B

Section C~C

Section D~D

Section E~E

Section F~F

Ridge line

Ceiling line

Floor line

Floor line

Floor line

Floor line

Floor line

Vertical section through party wall

Elevational section

Horizontal sections

Figure 10 *Common chimney arrangements*

It is common practice to re-line flues today with smaller diameter tubing to contain the gases from new appliances, but this does nothing to improve the fabric of the original construction. Frequently, when rebuilding chimneys at roof level, it is found that the withes are being restrained in position by the remains of pargetting only in the corners of the flues, and that the bedding mortar has crumbled and eroded away.

3.6.5 Rear extensions, bays, porches etc.

Rear extensions beyond the main body of a building were commonly added on towards the end of construction. There was a major advantage to the builder in maintaining easy access for materials to several levels of a building (e.g. by omitting bay windows and rear extensions until as late as possible in the works). More delicate decorative elements, such as porches and oriel windows, were naturally omitted until risk of damage during other construction was minimised. The foundations of these extensions were often at a higher level than the main building, and they were often on backfill, resulting in differential settlement.

These later additions would have had to be brick bonded or 'toothed-in' to the earlier construction. Apart from the difficulty of properly mortaring the joints to produce adequate bond, there is a small shrinkage associated with setting and drying which occurs early in the life of a mortar, and which might encourage cracks between new and older construction. In many areas, no attempt at brick bonding was made, and vertical fractures resulted.

The purpose of bay windows was to allow the maximum light into a room, and mullions were consequently made as small as possible. This restriction in width, together with the requirement of housing window box frames, afforded little opportunity for satisfactory construction. Stack-bonded piers with only nominal overlapping at corners were the norm. In some cases, upper bays were built off the timber window frames below. Concealed-brick relieving arches (Figure 11) often spanned over the bay window opening. The lintel might provide some tie action, but cracking from inadequate buttressing of the arch might have developed.

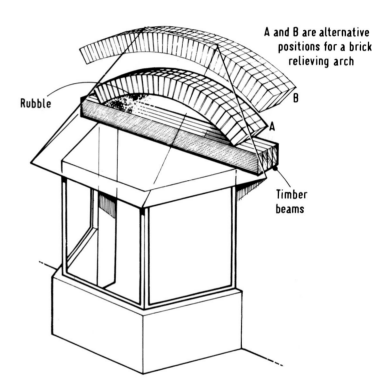

Figure 11
Relieving arch hidden by bay window

3.7   APPRAISAL OF EXISTING MASONRY CONSTRUCTION

3.7.1 Suitability for retention
The most common question to be asked at the outset of any refurbishment scheme is 'Can the existing structure be retained?' In many cases, this is a straightforward task, which could consist of listing the defects which require attention. The repairs can be costed, and after it becomes an accounting decision as to whether to repair or rebuild. Such cases are not the concern of this Report. Here, we consider cases where there is genuine concern as to whether retention is structurally possible because of the following factors:

1. Gross defects in whole structure
   foundation failure
   brickwork severely out of plumb
   extensive cracking.

2. Extensive local defects

    general delamination of brickwork

    extensive rotten bonding timbers.

3. Major alteration proposed

    large or numerous new openings

    increase in loads

    demolition of adjacent walls and floors affecting stability

    earthworks, excavations and changes in ground level.

The approach to the structural appraisal of buildings is presented in Section 2.3. An appraisal of the masonry elements should be carried out in the light of the following:

1. How does the masonry resist load at present (i.e. what is its thickness, out of plumb, loadings and load points, and where and how is it stabilised)?
2. How could the brickwork resist load as proposed?
3. Do the repairs and alterations improve its overall condition, or do they make it worse?
4. Is the structure subject to planning/conservation restrictions over rebuilding?
5. What is the extent and nature of previous alterations.
6. Are there any worthwhile spin-off benefits to be derived from rebuilding (improved insulation and damp proofing, longer life expectancy)?

Two drawings of the elevation of the wall should be made, showing it as a structure with its stabilising structures both before and after the proposed alterations. The site inspection work necessary to do this helps show up some of the defects, and it suggests where there may be others. In most cases, a basic appraisal can then be done by a qualitative analysis of the structure rather than by detailed calculations. A more detailed review of these factors in the retention of existing façades is presented in Section 7.

3.7.2 Ability to sustain increased loads

Brickwork may be required to sustain over-all, as well as local, increases in load. Over-all increases in load can occur as a result of adding new floors and increasing loads on existing floors.

In these cases, it is unlikely that a small general increase in brick stresses alone would be critical in long runs of wall, provided that existing gross defects are repaired and that there is no reduction in stability. Local stress increases in piers may need checking, and isolated piers of doubtful quality may need replacing.

The factor requiring closest examination is the increase in foundation loading (see Section 6). Over-all concentric load increases of 5 to 10 % usually prove acceptable on walls which have not previously been subject to increased loading, which have not suffered significant settlement, and where there are sound ground conditions as proved by trial pits adjacent to the footings. The means of foundation strengthening are covered in Section 6.2.

Most walls fail by some form of instability rather than crushing. Provision of adequate lateral restraint by connecting the wall to floors and crosswalls can greatly enhance its load-carrying capacity. This is often the best means of improving the integrity of the building.

Local increases in load, leading to an increase in brick stress, can occur because of the increase in load on an existing isolated pier, because new pier openings are formed, or because of local bearing stresses under new beams. Padstones may be used to distribute concentrated loads onto the brickwork.

A permissible compressive stress of 0.42 N/mm^2 (60 lbf/in.^{2}) has almost become traditional in many areas for old brickwork with lime mortar. This limit presumably derives from Table 3(a) of CP111 [3], and it relates to bricks with a crushing strength of 7 N/mm^2 (1000 lbf/in.^{2}) set in non-hydraulic lime mortar. In most cases, the time and expense of testing is not justified, and a stress of 0.42 N/mm^2 is a reasonable limit for the lowest standard of brickwork of the 18th and 19th centuries, where it is not affected by mortar decay. Tensile stress cannot normally be developed, and therefore the wall behaves as a gravity structure.

In larger projects where appreciable savings can be made if higher stresses are used, the testing of representative brick samples is a possible basis for determining design stresses, using the appropriate stress value in CP111 [3] for lime mortar. It may also be possible to test representative sections of brickwork cut from the building. Cement pointing can contain the lime mortar, and it can increase the compressive strength of the brickwork.

3.7.3 Factors influencing the stability of walls

The stability of a wall is influenced by many factors (e.g. the materials used, the form of restraint, openings and piers, and loading), so that an assessment by calculation alone should be supplemented by engineering judgement based on the following argument.

Consider the limit of stability of bricks loosely stacked and bonded as a free-standing wall. In fact, this model is not far removed from many walls of older construction comprised of bricks set in poor quality lime mortar. Acted upon solely by gravity, such a structure is found to retain a precarious equilibrium even when stacked to lean over by about 85 % of its thickness (see Figure 12(a) and (b)). Further movements first cause excessive local pressures, then over turning.

The provision of a relatively small degree of lateral restraint in the correct locations can render the wall capable of carrying additional load (see Figure 12(c) and (d)). Similarly, if the lateral displacements are limited to 50 % of the wall thickness, stable equilibrium is maintained under additional concentric vertical load without lateral restraint (Figure 12(e)). In this case, stresses are very high at the brick edge.

In a building, walls are more complex than loosely-stacked bricks. Even in masonry constructed with poor mortars, there is usually some adhesion or friction between the elements. Apart from free-standing garden walls, few walls are totally without any restraint, whether it is provided by returns at the ends, by the building in of a partition, or by a floor or beam. This membrane-like continuity explains why some grossly distorted walls still stand when simple mechanics suggests that the point of collapse has been passed.

3.8   ASSESSMENT OF DEFORMATIONS

3.8.1 Out-of-plane deformations

A wall constructed with sound materials properly bedded and bonded, well founded and restrained at sensible intervals, has great capacity for carrying load. Bedding mortar may compress, and local flaws may result in some spalling, but actual failure triggered by crushing of the components of a wall is seldom seen.

Bulging and leaning of a wall may have been built in, or may have subsequently occurred. If adequate lateral restraint is provided at intervals (say at floor levels and at the junctions with returns and internal partitions) so that further horizontal movements are restrained, even defects in other aspects of the wall should not result in major deformation of the wall. In many buildings, the lateral restraint is just not adequate. For example, floors built (but not tied) into a wall have allowed the wall to move laterally, or end returns and partitions which, although bonded into the wall, do not have sufficient tensile strength to restrain the wall.

A good example of this is the flank wall of a Victorian house, which is apparently restrained by front and rear walls and perhaps a spine partition, but not necessarily by floors. Such a wall might be four storeys high and an average of 1½ bricks thick. Restrained in position top and bottom, the wall has a slenderness ratio of 35, which is relatively high. Restrained only at its base, the effective height of the wall is doubled, and the slenderness ratio becomes 70. In this case, the buttressing walls are responsible for continued stability. The front and rear walls are largely perforated by doors and windows adjacent to the flank wall, and the soft mortar bonding of the spine walls or the timber partitions are insufficient to prevent outward movement.

The result is fractures in the scanty brickwork above door and window heads, and a vertical or diagonal toothed crack in the mortar joints of the spine wall. Collapse of such walls occasionally occurs, but stability can be maintained by the provision of adequate ties, even after deformations are quite advanced.

3.8.2 In-plane deformations

Deformations which occur principally within the vertical plane of a wall necessarily arise from vertical or horizontal forces and displacements. In-plane deformations are less complex than out-of-plane movements, and their causes are fewer and often more obvious.

The principal cause of in-plane deformation of walls is foundation settlement, most commonly arising from inadequate original sizing, or from moisture changes in the soil, or, occasionally, as a result of altered patterns of loading. Ground heave can also be a problem, where, say, mature trees and their roots are removed from a clay subsoil which slowly expands (and softens) as its moisture content increases.

Such movements result in characteristic crack patterns (Figure 13) which may vary in size across a wall surface. Indeed, it is often these crack patterns which provide the clue to the origin of the movement. However, settlement cracking should not be confused with cracking resulting from shrinkage (particularly of sand-lime bricks in more modern buildings).

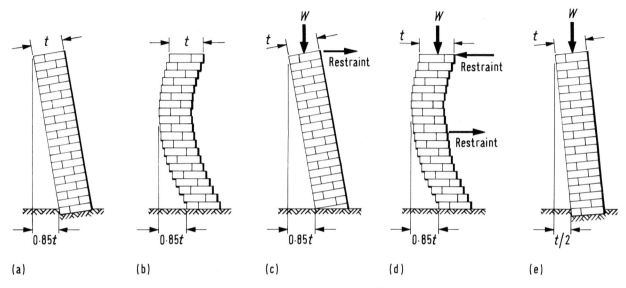

Figure 12 *Stability of free-standing and restrained walls*

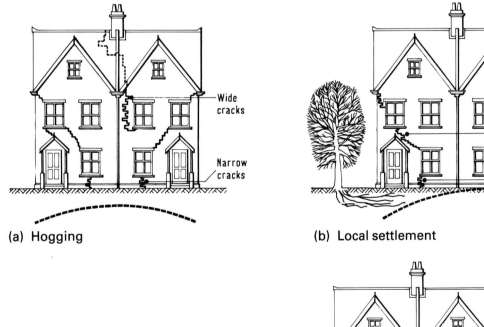

(a) Hogging

(b) Local settlement

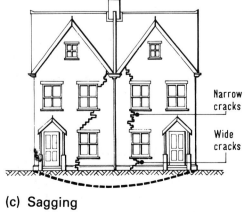

Figure 13
Cracking caused by foundation movement

(c) Sagging

In-plane displacements may occur where one wall relies upon another for lateral restraint. Often the buttressing wall is not strong or massive enough. Another common in-plane failure occurs where flat or low-rise arches are inadequately buttressed. However, the arch load rapidly finds an alternative load-path, so that, although the arch fails, it does not necessarily collapse.

From the structural point of view, planar movements are less serious than out-of-plane distortions, and they are commonly evidenced by sliding and opening of cracks. Most of the time, vertical loads still retain a fairly clear line of support or load-path down to the ground, and the occurrence of even quite large cracks, although detrimental to appearance and weather resistance, need not be structurally significant. Permissible crack widths are tabulated by BRE Digest 251 [14]. Crack widths less than 1 mm are termed 'very slight', and less than 5 mm are termed 'slight'. However, the Digest warns that these classifications relate to the degree of damage and whether or not the movements are active.

3.8.3 Survey of deformations

The lateral restraint required to stabilise a bulging or leaning wall increases as the deformation increases. The first step in assessing whether or not a wall can be retained is to measure its deformation by carrying out a plumb-line survey. Wall deformations can be extremely deceptive to the eye, and casual estimates made from ground or roof level are frequently incorrect.

In surveying a whole wall, not less than three plumb lines should be dropped, a good starting arrangement being one in the middle and one towards each end of the elevation. The positions of the lines should be adjusted or extra lines dropped to enable any large but localised distortions to be measured. It may also be necessary to measure the corner displacements. Readings should be taken at, and between, floor levels. It is most important to establish a base line from which the relative positions of all plumb lines are measured (Figure 14). In surveying walls with many windows, use of a theodolite or other optical methods may be acceptable¡ and easier to use than plumb lines.

In tall walls without windows, access for taking measurements can be extremely difficult. In these circumstances, it is often necessary to wait until a builder is on site and the wall scaffolded before the correct decision can be made regarding repair or reconstruction. There is usually no objection to this last-minute diagnosis, as long as all the options have previously been considered and a decision is forthcoming without delay after the detailed assessment has been made.

In assessing large displacements (say greater than one third of the wall thickness), it is important to know if the wall is still subject to movement (e.g. because of failure of its foundations, crushing of wall timbers, or thrusts from the roof timbers).

It is frequently not appreciated that deformations superficially of similar importance may have totally different significance. A deformation which is no longer active or is near an acceptable state of equilibrium need not necessarily give great cause for concern (e.g. the bulging of a wall subjected to some minor foundation movement which ceases after consolidation of the ground). A smaller, more local active deformation, such as a bulge arising through breakage of bonding stones in a rubble-filled wall, may be extremely serious if the deformation is accelerating towards collapse.

3.9   MEANS OF PROVIDING LATERAL RESTRAINT

A simplified analysis of over-all stability and a detailed examination of the buttressing structure should be made if retention of a wall is contemplated. In addition, calculation of the necessary restraint forces should be carried out so that ties and their anchorages can be properly sized. Provision of these or additional restraints can greatly improve the load-carrying capacity of the wall.

In accordance with normal structural design practice, the restraint forces may be based on 2½ % of the axial load in the wall, plus the outward component of force resulting from any out of plumb of the wall, and any wind forces.

Restraint may sometimes be provided in the form of external tie-plates bolted back through the wall and anchored internally to floors or partitions. The detail is a traditional one, but the decorative old cast-iron bearing plates are unfortunately hard to obtain. Bitumen-painted steel plates are now used.

In situations where no lateral restraint exists, a number of lighter gauge ties may suffice, cast into the inner face by concrete and again fixed direct to floor joists or suitable partitions. However, where significant load increases are proposed, more substantial ties may be necessary.

Where little lateral restraint is present at corners or masonry returns, either because the bond has been broken by the differential wall movements or because the original construction was not satisfactory, the restraint should be improved by the provision of steel angle straps or reinforced concrete elbow ties cast *in-situ* (say two or three per storey). The concrete would be cast into the brickwork using a shutter (the 'letter-box' method). When hardened, dry pack would be inserted to fill the gap and to enhance the bond between the concrete and the brickwork. Examples of these tie methods are shown in Figure 15.

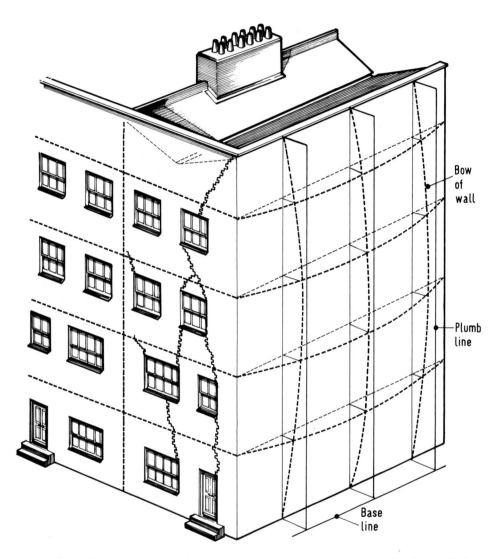

Figure 14 *Measurement of displacement of the flank wall of a building*

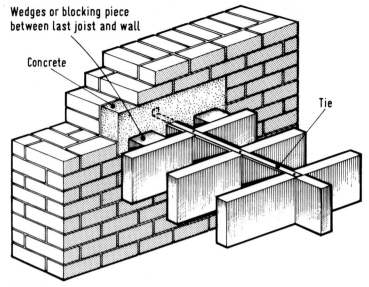

Wedges or blocking piece
between last joist and wall

Concrete

Tie

(a) Strap tie connection into wall

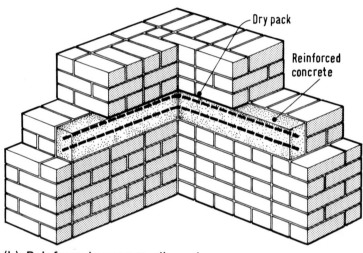

Dry pack

Reinforced
concrete

(b) Reinforced concrete elbow tie

Figure 15
*Examples of
providing restraint
to walls*

(c) Tie plate connection through wall

4. Timber construction

The majority of constructions which survive from the ancient world are built of stone, and it is often forgotten that timber has been used for building construction for at least as long. The development of timber in buildings has been a continuous process, reflecting its availability, the relative costs of material and labour, development in processing and distribution, and an increase in understanding of its structural behaviour and of the problems of rot, insect attack, and fire.

The period covered by this Report starts at a time when hardwoods had already become too scarce to be used in ordinary building structures. The problems of fire between buildings, and the long-term risks of rot and insect attack were not fully considered.

4.1 ASPECTS OF DECAY AND TREATMENT

It is mainly rot or insect attack which affects the serviceability of timber structures. When an attack is found, it is best to obtain the advice of a qualified timber specialist. He is able to identify the type of decay, advise on the treatment for the timbers or associated masonry, and identify those parts which should be removed. In the case of fungal decay, he should specify the cause and suggest measures for its rectification.

The design team should be prepared to consider the wider implications of the occurrence of the attack using the following criteria:

1. Was the incidence solely the result of a shortcoming in the original design (e.g. timbers without protection from moisture in ground, roofs with inadequate falls, lack of ventilation)?
2. Was it the result of some later modification (e.g. dampcourse bridged by soil or render, loss of ventilation)?
3. Was it the result of lack of maintenance (e.g. blocked or broken drains, or broken roof tiles)?
4. Do the same details exist elsewhere? Is there a possibility of further attack?
5. Is it reasonable to repair the defects, then to remove the potential cause of attack?
6. What are the best means of reconnecting new timbers to the affected timber?
7. What finishes have to removed to carry out the repair?

There are various chemicals for timber treatment, and although designers do not need to be experts in chemical formulae, they should be aware in broad terms of what can be done and how the work should be specified. Many of the preservatives for timber are listed in BS 5268: Part 5 [15]. A specialist sub-contractor may be employed, depending on the scale of the work.

4.1.1 Treatment of new timber

All new timber can be treated with chemicals which minimise the risk of rot (wet or dry) and eliminate the risk of insect attack. The most effective treatments are pressure impregnation with copper, chrome or arsenic salts. These are industrial processes carried out to order. The chemicals are waterborne, and the timbers take a long time to dry out again.

An alternative treatment, which is not quite as durable but ideal for joinery, is the double vacuum organic solventborne method which avoids drying out problems and dimensional changes. These treatments drive the chemicals into the timber to a much greater depth than surface application by spray or brush. Even so, the timbers are not treated right through, particularly in heart wood, so that notches and ends cut after treatment need to be dipped or brush treated on site. If this is not done, the protection is by-passed.

4.1.2 Treatment of existing timbers

The forms of attack in timber are described in detail in BRE Digest 299 [16] and BRE Technical Note 44 [17]. The main forms of rot and attack are:

Wet rot

Excluding moisture is all that is required to stop wet rot spreading. The strength of the timber section may well be reduced, requiring additional strengthening when it has dried out. There is also a risk of dry rot breaking out as the timbers lose moisture. This risk increases as the timbers pass through the ideal moisture content for the inception of dry rot (20 to 40 % weight/weight), and it ceases during the final drying to ambient conditions (10 to 16 % weight/weight).

Dry rot

An unchecked outbreak of dry rot is extremely serious, and its eradication requires drastic treatment. All infected timber must be removed, and it may be advisable to remove a further 500 to 1000 mm beyond the extent of the apparent infection. This major surgery can have a significant effect on the structural action, and it needs careful thought on how to replace the cut members to minimise the disturbance to the whole structure.

It is vital to determine the extent of infection. Once established, dry rot strands are quite capable of passing behind plaster and through brickwork until they reach other timbers which may be far distant from the primary moisture sources. These can then be attacked with the same consequence as at the source of the outbreak. Once all affected timbers have been identified and removed, the remaining timbers may receive preventative treatment. It is also advisable to sterilise brickwork within and around the affected zone. Surface treatment, and sometimes drilling to the perimeter of the affected area, are usually all that is required. Mass irrigation by drilling holes at say 300-mm centres over a wide area can be damaging to the structure of an old brick wall, and it should not be considered necessary in most cases. Treatment with a blowlamp is both ineffective and hazardous.

Insect attack

Where there is evidence of recent insect attack, existing timbers should be treated on their surface with an insecticidal fluid. A waterborne treatment tends to penetrate about 2 mm, and it is only really effective against emerging and egg-laying insects. An organic solvent-borne treatment penetrates up to 15 mm, and it is much more effective for long-term protection.

There are also chemical pastes for application to buried and large section timbers. These release chemicals gradually, achieving deep penetration over a period of time. Up to 50 mm penetration into the side grain has been observed in softwoods. Treatment of large timbers can also be assisted by drilling and inserting chemicals into the centre of the section.

4.2   ASSESSMENT OF STRENGTH

Whenever an old structure is being retained or restored, the question of estimating the strength of the structural timber elements is important. The strength and elastic modulus of old timbers are often significantly greater than those given in CP 112 [18].

However, the apparent capacity of the members in terms of the timber sizes, spans and strengths is seldom matched by the strength of connections. It is also important to consider whether deflection may be the governing criterion, particularly where long-span timber or flitch beams are concerned. There may also be a major disparity between the capacity of the primary and secondary beams in a floor.

The following information should be collected:
1.   identification of the species and quality of the timbers (including extent of infestation or decay)
2.   determination of member sizes, and their over-all geometrical relationship
3.   examination of joints and connections
4.   existing deformations.

In extreme cases, a load test may be required. Items (2) to (4) are within the normal expertise of the engineer, but the positive identification of timbers is a specialist task.

The Timber Research and Development Association operates a service on a fee basis which can provide this information. Species can be identified by supplying them with a small (but not too small) sample. However, visual stress grading can only be done on site. Such inspections are easily arranged, and they are successful when the timbers are fully exposed (e.g. in warehouse buildings without soffit ceilings), but less so in cases where inspection is limited. This limitation arises because the 'grade' of the timber is related to the worst degree of wane, knots, or notches which are present.

4.3   TIMBER FLOORS, ROOFS AND PARTITIONS

4.3.1 Timber floors

The layout of timber members in floors can differ considerably with the age of the building. In many larger buildings of the 17th and 18th centuries, heavy timber beams were used, onto which secondary beams and joists were placed. These joists were often cut or 'cogged' onto the main beams, to minimise the floor depth, the main beams often being the weakest elements.

Timber floor joists were available in lengths far greater than could be easily obtained today, and joists running from front to rear of large houses are quite common. This arrangement often allowed for the line of the load-bearing spine wall to be offset between one level and another.

The span direction and sizing of the floor joists can vary between apparently similar buildings and even within one building. In small Victorian buildings, the floor joists were commonly of 7- to 10-in. (175- to 250-mm) depth. The areas where the framing system was modified were at fireplaces and at bay windows. At fireplaces, a trimmer beam was used onto which sat the brick arch or the stone slab of the hearth (Figure 16). The trimming joists which supported the trimmer were commonly the same depth but thicker. It was normal for rows of herringbone strutting to be used to restrain the joists laterally at certain points (typically at one-third points along the span), and to distribute concentrated loads.

4.3.2 Timber trusses

The high strength to weight ratio of timber when stressed parallel to its grain makes it an ideal material for truss members, providing sections free from gross defects are chosen. In timber trusses, the member sizes are usually governed by the design and detailing of the joints. This differs from concrete and steel design, where sections are usually determined by stresses within a member.

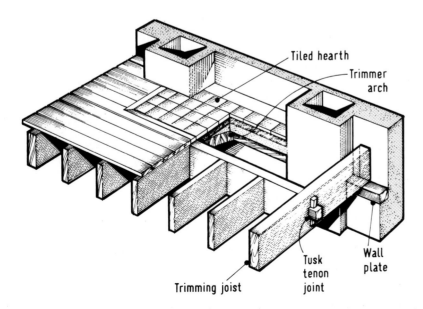

Figure 16 *Example of framing at fireplace*

Unlike steel trusses which use many small members and connections, older timber trusses usually consist of a small number of principal strut and tie members with interlocking connections and wrought-iron straps. Loads from often substantial roof purlins are usually applied close to truss connection points. These trusses differ markedly from modern TRADA-type roof trusses, which use many bracing members with metal connectors, and which are laterally very flexible.

4.3.3 Trussed partitions

Before the latter part of the 19th century (when rolled iron or steel beams became readily available), it was usual for loads to be carried over large openings linking principal rooms by means of trusses built into the partitions in the floors above or below. An example of such a trussed partition is shown in Figure 17.

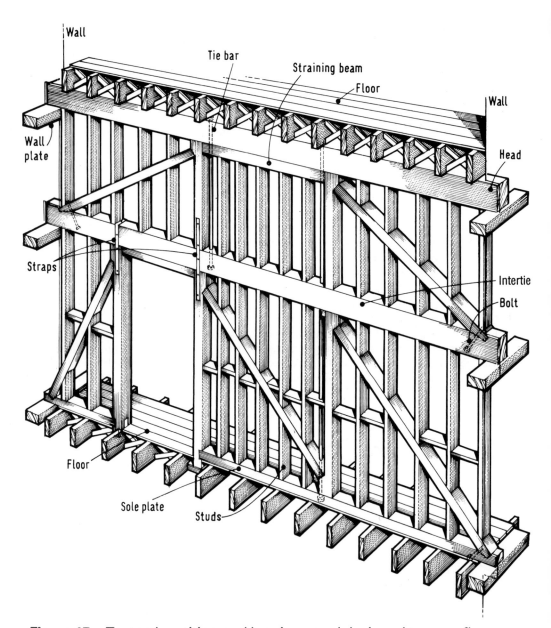

Figure 17 *Trussed partition and bracing panel designed to carry floor load*

Even the simplest stud wall without openings was usually braced by a diagonal stud. This enabled 'tight' construction to be achieved, because the studs did not need to be cut to exact length. Many ordinary partitions between rooms, as in Figure 18, may also give the appearance of a truss, and may therefore appear to have some load-carrying capacity. However, it is unlikely that any tying action exists where the partition runs across the main floor joists and is perforated by original door openings.

There may be other cases where the removal of diagonal bracing panels beneath a trussed partition causes the truss to deflect further, because of loading on the partitions resulting from long-term creep of the timbers above. Such 'semi-structural' situations are commonplace in this form of construction, and their action is impossible to deduce without producing a full elevation sketch.

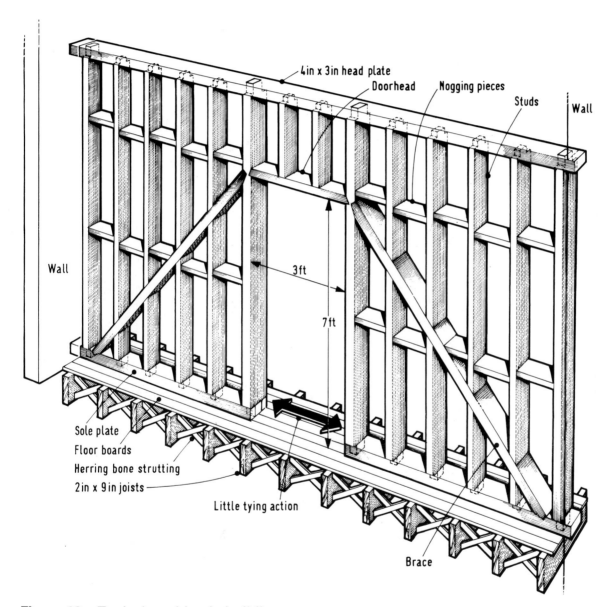

Figure 18 *Typical partition in building*

4.4 TIMBER CONNECTIONS

In modern timber construction, all but the simplest joints are made with the aid of metal connectors: screws, framing anchors, joist hangers, and toothed plates. In the period covered by this Report, metal connectors were unusual, except in such elements as flitch beams and roof trusses.

In other joints, the loads were generally transferred by direct timber to timber contact, the wood having first been formed into quite complicated shapes to very small tolerances. The assembled joint was often nailed or dowelled to prevent the pieces coming apart, but the nails do not directly contribute to the load capacity of the joint. Typical connections are shown in Figure 19. Once assembled, the true shape of the timber contact areas is not visible. A knowledge of typical details, and a careful inspection is therefore needed before the likely load-carrying capacity of such joints can be assessed.

In a typical floor framing plan, it is common to find 'morticed' or 'housed' joints between the joists meeting at right angles with no other form of fixing. 'Tusk-tenon' joints Figure 19(g) may be found in the trimmer beams at fire places (see Figure 15, page 36) where the 'key' was used to resist any tendency for the members to separate.

Truss connections are generally tenoned and dowelled. The joints between the strut and tie members (Figure 19(a)) are required to transmit shear and compression to a support. A bolt through, or strap around, the joint is used to prevent separations. A 'king post' connection (Figure 19(f)), despite its size, normally only resists tension. A metal stirrup tightened by cotter-pins or wedges is used to support the beam at this point.

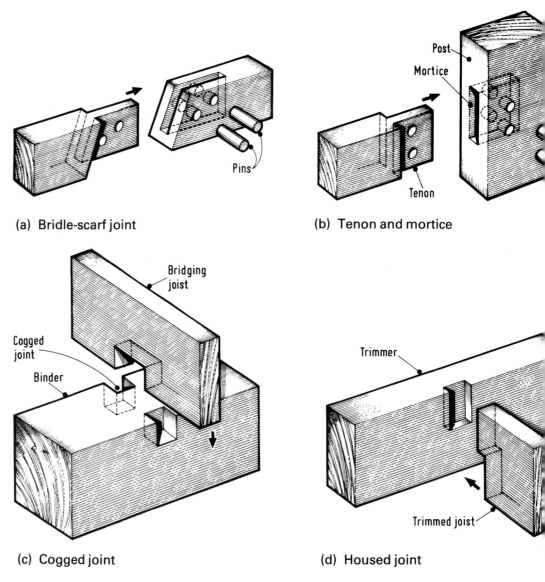

(a) Bridle-scarf joint

(b) Tenon and mortice

(c) Cogged joint

(d) Housed joint

Figure 19 *Forms of traditional timber joints*

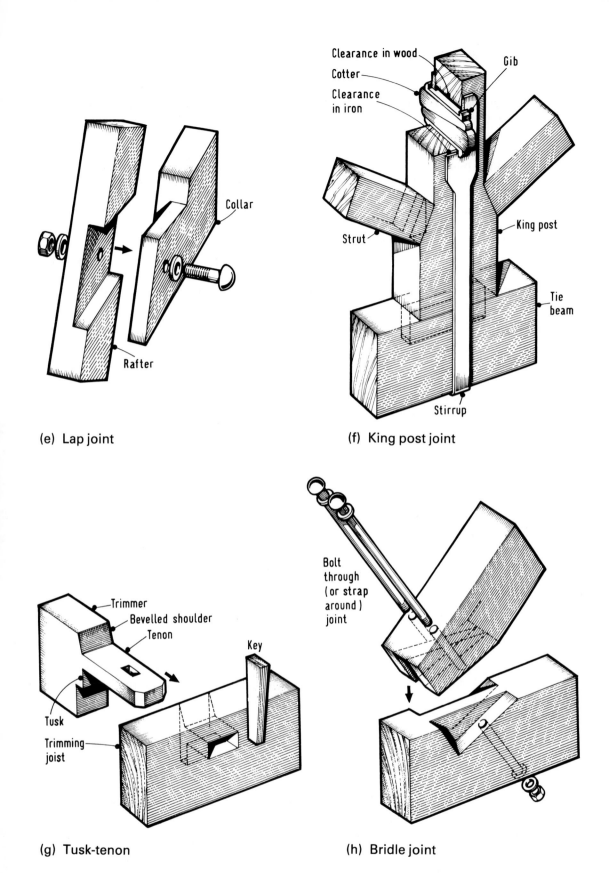

(e) Lap joint

(f) King post joint

(g) Tusk-tenon

(h) Bridle joint

Figure 19 – *continued*

4.5   COMPOSITE OR SPECIAL TIMBER MEMBERS

Timber is used in many circumstances in a form where it acts compositely with itself or with other materials. There are some examples which may be considered to be traditional features within buildings.

4.5.1 Flitch beams

Flitch beams are formed by taking a solid timber beam (usually about 300 to 400 mm square) and dividing it in half longitudinally. The two halves are then reversed, and one section turned 'upside down' (see Figure 20). The result of this is to ensure that any natural defect which may have occurred originally is divided, and that it no longer dominates any one section. Higher strengths are therefore possible than for the original timber.

It is common for a steel or iron plate, of similar depth to the timbers, to be sandwiched between the timbers, and the whole member to be bolted together. Other forms of intended strengthening or stiffening (e.g. by strips of hardwood or iron) may be hidden between the two sections of a flitch beam to form a trussed girder [19]. Although this latter method may not contribute to a significant increase in strength, they suggest an awareness by the original builder of the need to maximise strength by eliminating material weakness, and they are symptomatic of higher quality construction.

4.5.2 Relieving arch

Timber beams may apparently support loads from walls above, but they are often relieved of a large proportion of their load by masonry arches built into the wall. These 'relieving arches' were sometimes hidden externally by roofs over bay-windows. However, the load resistance is a composite action, depending on the position of the arch in relation to the timber beam (see Figure 11, page 30).

4.5.3 Bressummers and storey posts

The term 'bressummer' was first used for the first floor beams in the external walls of medieval timber-framed houses, but subsequently it denoted any beam supporting brick walls over a wide opening. Bressummers were often used in industrial buildings and in Victorian shop-fronts. In this latter case, they may have been inserted into the existing construction, and the narrow pilasters on shop-fronts were such that the ends of the timber bressummers may have had inadequate bearing onto slender timber 'storey posts' adjacent to the party walls. Intermediate storey posts were sometimes used to support the bressummer and the wall above.

Problems often developed where the bressummers and storey posts were inserted later, because of creep deflections, and crushing or rot of the timbers. In later Victorian buildings, iron replaced timber. Where these elements were built into the original construction, problems would have been less likely, but even so, any defects and deterioration should be identified, particularly if any increase in load is proposed.

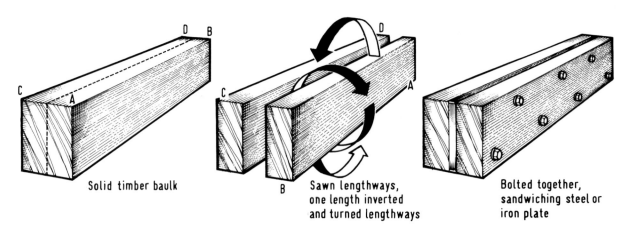

Solid timber baulk   Sawn lengthways,   Bolted together,
                            one length inverted   sandwiching steel or
                            and turned lengthways   iron plate

Figure 20   *Flitching of timber beams*

4.6 REPAIR OF TIMBER

There are many reasons why it may be necessary to repair or replace a timber member. Examples are indicated in Figures 21 to 24, together with suggested methods of repair.

Posts rot at their bases, and trusses or joists at their wall supports. Cutting back and replacing the affected timber, taking the appropriate measures to eradicate the source of the decay, is often the only solution. A lead damp-proof course and packing below timber posts or other supports is a good precaution against moisture.

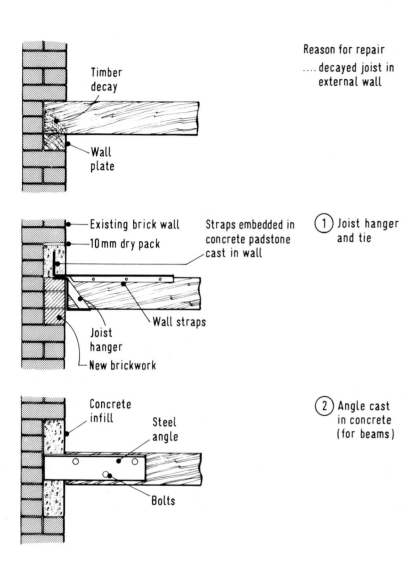

Figure 21 *Examples of repair of decayed joist at wall connection*

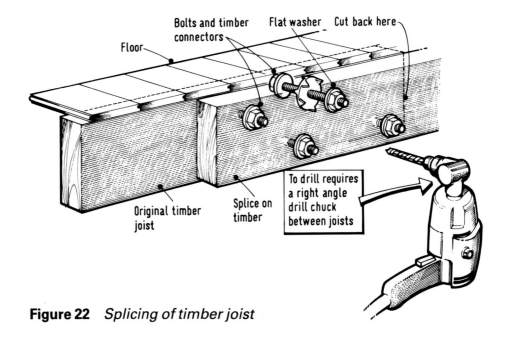

Figure 22 *Splicing of timber joist*

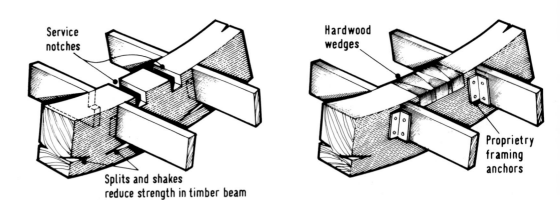

Figure 23 *Repair of notched beam*

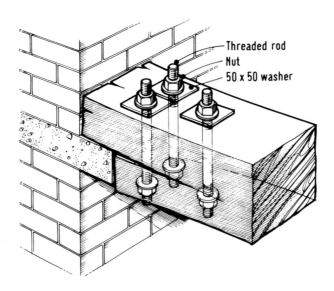

Figure 24 *Strengthening of split timber*

4.6.1 Repair of beams

A common problem shown is that of decay at a 'wall plate', which may be resolved by cutting back and supporting the floor beam by a bracket or by a joist hanger (see Figure 21). The decayed wall plate should be removed. The relative merits of the repair methods indicated depends on the ease of access, the quality of the brickwork, and the size and spacing of the timbers to be repaired.

It may also be necessary to splice two members together by bolts and modern timber connections. This form of joint may be able to transmit the required moment close to a support, but it rarely develops the full flexural capacity of the members. The practical difficulty of splicing two joists is increased when they are close together, and a special right-angle drill is required to form the holes (Figure 22).

Old timbers may have been notched, either by means of jointing or by channels cut for more recent services. This may cause excessive deflection or, in extreme cases, failure under load. It is often possible to fill in the notches, using hardwood wedges as shown in Figure 23. This does not cause any reduction in the existing deflection, but it stiffens the floor to imposed load.

Where splitting of a member has occurred close to its support, it may be possible to retain the integrity of the member to shear by using bolts through, or straps around, the section as shown in Figure 24. Such splits and shakes are common in old timber, and a decision has to be taken as to whether they are structurally important or not.

Strengthening of timber beams may be possible by splicing steel plates onto the webs, or where accessible, onto the top and bottom to form flanges. The number of bolts or studs should be sufficient to develop the required force in the steel. Providing secondary timbers to relieve the overloaded joists is also a simple solution.

4.6.2 Repair of trusses

Trusses, in particular their connections, can fail in a variety of ways (Figure 25). In tension splices, it is usually the crushing strength of the timber close to the connector or pin, or failure arising from insufficient edge distance which controls. At the joint between the principal rafter and the tie, it is the crushing or shear strength of the timber which controls. The bearing strength of 'scarf' connections is usually quite considerable, although bolts or straps contribute little to the joint strength.

Where decay occurs in trusses close to their supports, it may be possible to splice new timbers after removing the affected section. A gusset plate may be appropriate for repairing a truss, comprising smaller timber members. Examples of these methods are shown in Figure 26.

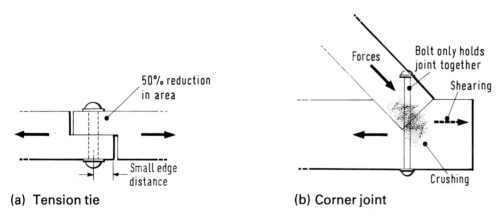

(a) Tension tie   (b) Corner joint

Figure 25 *Points of potential weakness in timber trusses*

4.6.3 Epoxy resins

Developments in polymer chemistry have led to a range of specialist repair techniques using epoxy resins. The characteristics are that resins of low viscosity and surface tension are able to permeate and saturate partly decayed timber. On hardening, the resin/timber combination acquires properties more akin to the cured resin than to timber. It is also possible to achieve a good bond to stainless steel rod, so that badly decayed timber can be drilled, stitched and glued to achieve higher strengths than the original construction. However, this is a very expensive process.

Epoxy resins are particularly appropriate to the repair of failed joints on timber framed structures, where the removal and replacement of individual members is impracticable. Other uses include the *in-situ* repair of principal timber floor beams without removing floor and ceiling joists. Special consideration should be given to the performance of epoxy resins in fire.

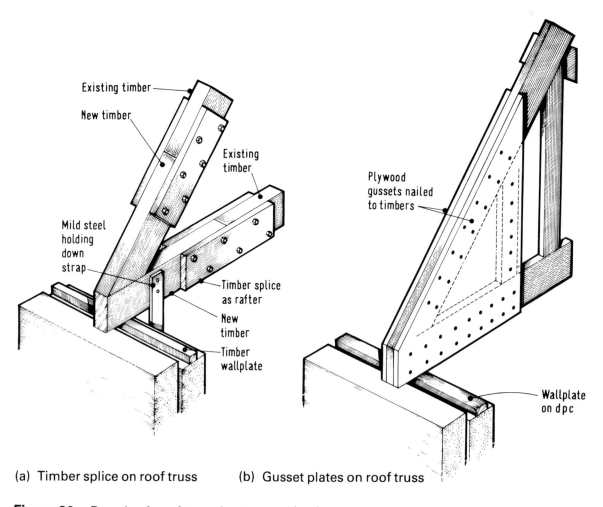

(a) Timber splice on roof truss   (b) Gusset plates on roof truss

Figure 26 *Repair of roof truss by two methods*

4.7   STRUCTURAL TIMBER IN FIRE

Building Regulations dealing with timber in buildings do so in terms of three distinct requirements in fire:

1. 'Surface spread of flame'. This is tested to BS 476 [20], and controls are imposed to prevent fires spreading from one part to another, and to ensure that escape routes in particular are not lined with flammable materials.

2. 'Non-combustible' construction. The word 'construction' is chosen carefully here, because it covers more than just the structure. For example, non-structural timber stud walls are controlled under this category. The aim of this control is to limit the amount of fuel incorporated in the building structure, which could feed a fire established in some other way.

3. Structural fire resistance. Under this category, structural elements are tested to BS 476: Parts 20 and 21 [20]. The structure is required to withstand, under load, a period of exposure to heat without collapsing and remain within certain deflection limits. It has also to be capable of being re-loaded 24 h later.

It is important that these three categories are understood, for while exposed timber surfaces can be treated to reduce surface spread of flame, and the structural fire resistance of timbers can be justified by analysis, the problem of 'incombustible construction' can only be resolved by a relaxation of the regulations, justified by measures to reduce the risks.

4.7.1 Structural fire resistance of timber

When subjected to gradually increasing temperature, timber reaches a point where charring occurs. The presence of flame is not necessary.

The two main factors which influence the fire resistance of timbers are:

1. Within the temperature range of up to 1-h tests, the rate of charring is constant and predictable.
2. Timber suffers no appreciable loss of strength until charring occurs, so the full strength of the uncharred parts is always available.

Thus, at any time, the net section remaining can be predicted, and the strength of the member assessed, using the higher stresses permitted in fire. This method is explained in BS 5268: Part 4 [21].

Special consideration should, as always, be given to the joints of a timber structure in a fire. Sections of reduced cross-sectional area may become completely charred away in advance of the main member, and unprotected metal parts either fail (e.g. framing anchors) or introduce heat and charring into the centre of members, increasing the number of directions in which charring advances (e.g. where a steel plate is sandwiched between the timbers in a flitched beam).

In situations where the structural fire resistance cannot be determined using this method, additional fire protection has to be applied. Building control regulations contain tables of 'deemed-to-satisfy' methods, but these are chosen as appropriate to common forms of new construction. The manufacturers of plaster and applied fire protection specialists may often have unpublished BS 476 test results for their materials in uncommon situations. Intumescent materials, which expand on heating, are not generally appropriate, because the temperature at which they intumesce is similar to that at which timber charring occurs.

4.7.2 Fire protection of timber floors

Timber floors are classed as combustible construction. However, there are various means by which their structural fire resistance can be improved to satisfy the requirements of BS 476: Parts 20 and 21 [20]. It is often possible to obtain a relaxation to any incombustibility requirement. Any upgraded floor also has to satisfy the integrity and insulation requirements of BS 476 so that the fire does not spread from one floor to another.

The principles which can be used to satisfy these requirements are: sacrificial timber (allowing for charring), applied fire protection, and redundancy in the event of fire. Examples of these methods are shown in Figure 27. In an unprotected floor, the limiting factor is usually the thickness and integrity of the floor boards.

Examples 2 and 3 show two forms of applied fire protection. Example 4 has mass concrete 'pugging', which provides additional stiffening and sound insulation as well as enhancing the fire resistance of the floor. The joists and their supports need to be generously sized to carry this extra dead load. Other types of sound-reducing pugging, such as sand, make no contribution to the fire resistance, and some, such as sawdust, may be hazardous in the event of a fire.

The design of a new concrete floor to carry the imposed loads can be a useful method where the 'joist effect' is to be retained and where there is adequate headroom. The joists have to be able to resist the self weight of the concrete floor, including an allowance for the loads during concreting.

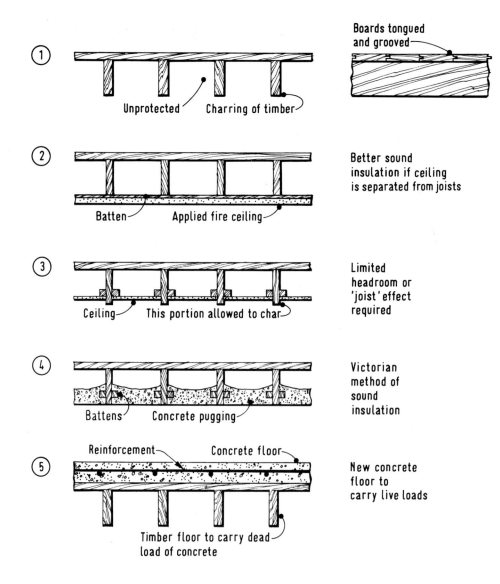

Figure 27 *Examples of fire protection of a timber floor*

5. Cast and wrought iron and early mild steel

5.1 HISTORY AND DEVELOPMENT

Cast and wrought iron were first used in buildings as ancillary and decorative elements, but their widespread use as structural materials came about as a consequence of the industrial revolution, and ended with the widespread availability of steel. The first major engineered or structural use of cast iron was the Iron Bridge at Coalbrookdale in 1779.

Early cotton mills, with traditional timber structures, suffered from frequent fires. 'Fire proof' (or more correctly non-combustible) construction was a significant development in the early 1800s, using cast-iron pillars and beams to support brick jack-arch floors (Figures 3, page 14, and 28). The iron technology came from the foundries which already produced castings for the mill machinery, and the huge engines used to power them.

One of major uses of cast iron around 1840 was in the construction of large warehouses (St. Katherine's Dock, London, and Albert Dock, Liverpool, are examples of recent renovation projects). Wrought iron transferred from the railway industry, where its development was encouraged by the need to construct longer span bridges and many miles of rails. Principal among these were the Conway and Britannia bridges whose construction in the 1840s was one of the most significant advances in structural engineering to take place in the 19th Century.

From the 1840s, the basic characteristics of the materials were established, although their quality varied considerably. Wrought iron was initially relatively expensive, and cast iron had the advantage that it could be moulded in decorative shapes.

The limitation of cast iron was its low strength/weight ratio, particularly in tension. To overcome this problem, cast-iron beams were cast with much wider or thicker bottom flanges. Various composite cast-iron/wrought-iron girders were experimented with, but not with any great success.

Wrought iron was produced by hammering and rolling from individual billets of iron, the weight of which was often limited by handling. This tended to limit the size and length of the early wrought-iron members. To overcome this, wrought-iron beams were also made up from plates or made into trusses to cover longer spans. Rivets were used to connect the parts.

Figure 28 *Jack vaults between wrought-iron beams*

Attempts to make wrought iron economically in large quantities were dramatically successful, but their success took the industry off in another direction. The Bessemer process, patented in 1856, set out to produce iron, but produced not iron, but steel, at about six times the speed of the iron puddling process. Initial samples of steel made by the process were unpredictable and prone to brittle failure. However, the quality of mild steel soon improved, and its allowable strength was greater than wrought iron. However, it was some 20 years before steel was produced in quantity with reasonable quality control. One of the major engineering achievements in steel was the Forth Bridge, completed in 1890.

Steel was first used in the construction of the expanding railway system. The first rolled-steel joists were produced in the 1880s, and from then on this material gradually took over from cast and wrought iron, the process being completed by the 1920s. A recent BCSA publication [22] gives comprehensive data on old I sections.

5.2 IDENTIFICATION AND FORM OF IRON STRUCTURES

The main characteristics and form of cast-iron, wrought-iron, and steel structures are summarised in Table 2. Identification is aided if the age of the building is known, although between about 1860 and 1890 all of the materials were in use. The different forms of iron used in beams are summarised in Figure 29. Cast-iron beams are noticeable by their heavier bottom flanges, and their shape was often determined by their function. For example, 'springing' members in the form of inverted T or V shapes were used to support masonry arches. Mild steel sections only came into use in the latter part of the 19th century, after which time the use of cast and wrought iron declined rapidly.

Cast-iron columns are usually of circular or cruciform shape with different forms of head detail. In warehouse construction in the south, they were commonly used to support large timber beams as shown in Figures 30 and 31. Wrought-iron beams, large jack-arch or vault floors were more likely where heavier loads were to be supported, or greater column free spans were required. For example, in many of the large mill buildings in the north, all-iron structures were used.

Figure 32 shows examples of early 'fire-proof' floors. The scale of jack-arch construction varies considerably from long-span arches, of 2 to 4.5 m (Figures 3, page 14, and 28, page 51) to the small scale as in Figure 33. Filler joists generally used wrought-iron I beams of shallow depth. A common characteristic of wrought-iron and steel sections was that beams deeper than about 300 mm were made up by rivetting plate sections together. Wrought-iron and steel sections may be difficult to differentiate, apart from where the steel maker's name is indicated. Wrought iron may suffer from delamination when corroded, because of the presence of slag between layers of almost pure iron. Wrought and cast iron, and, later, wrought iron and steel may well be found in the same building, the more expensive wrought iron being used for roof trusses or tie rods, for example.

text continues on page 56

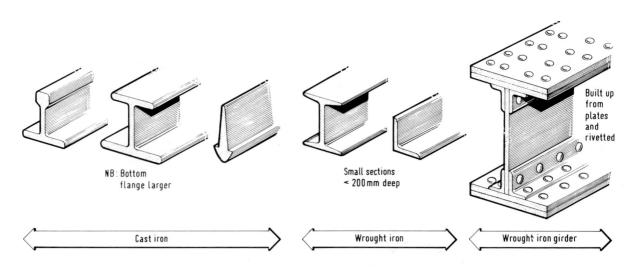

Figure 29 *Typical iron beam sections*

Figure 30
*Cast-iron
columns of
circular shape*

Figure 31   *Typical cast-iron columns of cruciform shape*

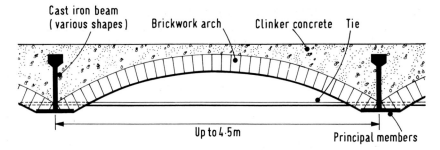

(c) Jack-arching between principal members

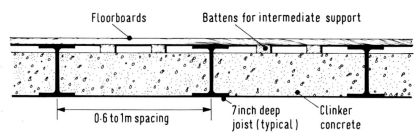

(b) Filler joist floor

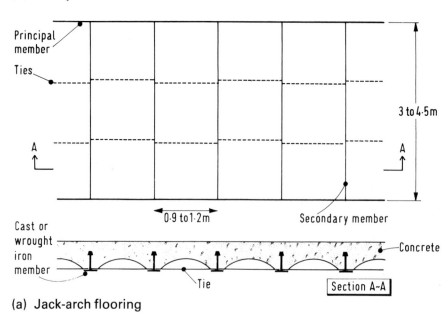

(a) Jack-arch flooring

Figure 32 *Early forms of iron floor*

Figure 33 *Typical jack-arch floor*

Table 2 *Characteristics of cast and wrought iron and mild steel*

	Cast iron (see Section 5.3)	Wrought iron (see Section 5.4)	Mild steel (see Section 5.5)
AGE OF BUILDING	Cast-iron pillars up to 1914	Rivetted girders unlikely before 1850	Mild-steel rolled joists after 1880
	Cast-iron beams unlikely after 1850 except jack arches or ornamental structures	Wrought-iron beams unlikely after 1890	Dorman Long section handbook in 1887
FORM	Cast-iron stanchions of cruciform or hollow section, recognised by end plates and decorative details	Small sections initially	Bolts, bars and I beams
		Built-up sections by rivetting later or in trusses	Built up sections
	Beams have larger bottom flanges and may be bellied on plan	Rods in jack arch floors	
SURFACE TEXTURE	Sandy texture when unpainted. Signs of mould joints. 'Blowholes' on one face may indicate open face. Corners rounded internally	Similar to steel, except more rounded corners	Smooth with sharp corners
DEFECTS	Cooling cracks. Uneven thickness of hollow sections. Surface pitting and blowholes (perhaps disguised)	Lamination if badly rolled	Fewer defects, because of better quality control
CORROSION	Good corrosion resistance	Moderate resistance	Poor corrosion resistance
FAILURE MODE	Brittle tensile cracking	Yielding or fatigue or rivet failure	Yielding or rivet failure

5.3   CHARACTERISTICS OF CAST IRON

5.3.1 Material structure

Cast iron is a crystalline material whose main constituents are iron and carbon. Most Victorian cast irons are 'grey' irons, and these are combined as a mixture of iron/carbon compounds and, usually, free carbon in the form of graphite flakes [23]. It is the free carbon which gives rise to the characteristic brittle behaviour and the low tensile strength of cast iron, because the presence of graphite flakes with negligible strength produces internal stress concentrations. Modern cast iron is a much stronger and more ductile material than that encountered in old buildings.

5.3.2 Ultimate strength

The strength of old low-grade cast iron is notable for its variability. In 1900, Twelvetrees[24] quoted average values for the ultimate (or breaking) strength of cast iron as:

	$tonf/in^2$	N/mm^2
Tensile strength	8	120
Compressive strength	38 to 50	590 to 780
Flexural tensile strength	15	230
Shearing strength	6 to 13	90 to 200
Elastic limit	1	15
Young's modulus		
compression	5570 to 5880	84 500 to 90 500
tension	4260 to 6070	66 000 to 93 500

The lowest strength of a series of tests may be some 30 to 50 % below these average values. It is clear that the factors of safety applied to these ultimate strengths to obtain permissible working stresses should be very high in order to allow for the variability of the material. An average factor of safety quoted by Twelvetrees was between 5 and 8, depending on the member. Cast iron is therefore more similar to a material like concrete where the average strength of grade 30 concrete might be 40 N/mm^2 and its permissible stress in compression might be 8 N/mm^2.

5.3.3 Casting imperfections

Imperfections derived from the casting process, rather than from the material. These are not unfamiliar to the concrete engineer – but at first sight may not be expected in a metal.

The material often set with extraneous materials or air voids contained within it. These were usually apparent (though not necessarily their extent) as soon as the mould was broken. As with modern structural concrete, if such a defect were too extensive, the member was recast. Minor defects could be filled (usually with lead).

The material entered the mould in a molten state, and set as it cooled. The temperature at which it set was well in excess of that at which mild steel is rolled. Any cast-iron member therefore underwent a reduction in size after it had solidified and as it cooled to ambient temperature.

For accurate casting, the mould pattern had to be slightly oversize, and the element had to be designed to avoid sharp changes in thickness. All corners had to have fillets or gentle radii. If these requirements were not met, the member cooled unevenly, and the result was at best the setting up of undesirable 'locked-in stresses' and at worst cracking of the member. When assessing the overall quality of an existing cast-iron structure, it is important to assess to what extent these factors were understood by the original designer.

The casting process made it possible to create hollow members – most notably hollow circular columns. This was done by placing a separate core piece within the main mould, supported at its ends, or at other points where holes could be permitted in the final member. In general, this meant that the degree of support to the core had to be kept to a minimum. Sometimes this was not enough, and the core displaced as the hot metal entered, giving rise to the variation in wall thickness typical of hollow cast-iron columns.

At first sight, this might be thought to be a serious defect, but in properly designed structures where the columns are not subject to eccentric loads of any great magnitude, variations of 6 mm or more in a 25-mm wall thickness are not significant. This is because:

1. The methods of analysis and safety factors used in the 19th century were based on tests on real columns, which had such defects.
2. Provided the core is only displaced (and not locally enlarged), there is no change in the net cross sectional area, and the radius of gyration of the section is not significantly affected.
3. There is a tendency for cast iron to be stronger if it cools quickly. The thinner wall sections tend to cool quicker than the thicker walls, and to be marginally stronger than the thicker parts.

5.3.4 Connections

As the casting process itself placed no limit on the size of cast-iron members, the need for joints was dictated by the design and proposed construction of the building. Joints most commonly occur at beam/column connections. Individual beams and columns were commonly cast in one piece, unless the weight of the component, or the intricacy of its shape made this impracticable.

There are three principal types of joint which may be found in structural cast iron.

Locating connection

This is by far the most common. It refers to a joint where the loads are transferred by direct bearing between one member and the other. The bolts or screws are provided to locate and stabilise the members, but they do not carry the principal loads. Examples are the bolting down of beams onto column heads, and the bolting together of column bases and heads. Typical beam - column connections are shown in Figures 34 to 36. In columns, the spigot and casting within the beam depth may be separate from both column members.

Sub-assembly connections

These occur in members where their shape makes a single piece of casting impossible, such as because of the nature of the decoration. These joints may be difficult to detect, and they usually take the form of countersunk wrought-iron screws. The expansion created by modest corrosion can lead to failure of the screws.

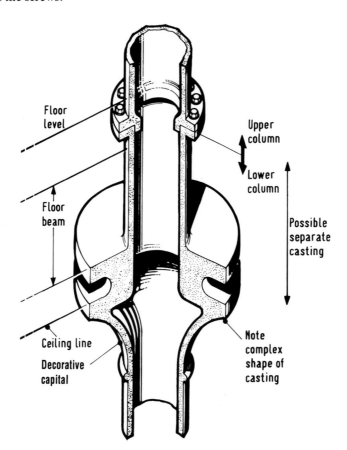

Figure 34
Detail of column head

Principal connections

In these, all the stresses at the joint in question have to pass through the screws or bolts. They are introduced to limit the maximum component weight, and they generally occur in long-span structures such as bridge beams. In buildings, the spans are shorter, and single castings were generally used.

Other methods

There are no satisfactory ways of forming structural connections in cast iron by any processes such as welding, brazing or soldering. If such processes appear to have been used, the structure should be regarded as suspect, because all informed Victorian engineers were aware of the risks of such connections.

Another practice to be regarded with suspicion is that of 'burning on'. This was carried out as a cheap alternative to breaking-up and re-casting defective castings. It was used, typically, where the majority of the casting was good, but where integral lugs, eyes, cleats, etc. were found to be cracked. The process involved forming a pattern only for the portion to be replaced, and fixing this to the main casting. The area of the join was then heated, and new molten metal was poured into the pattern. The pre-heating, and the heat from the molten

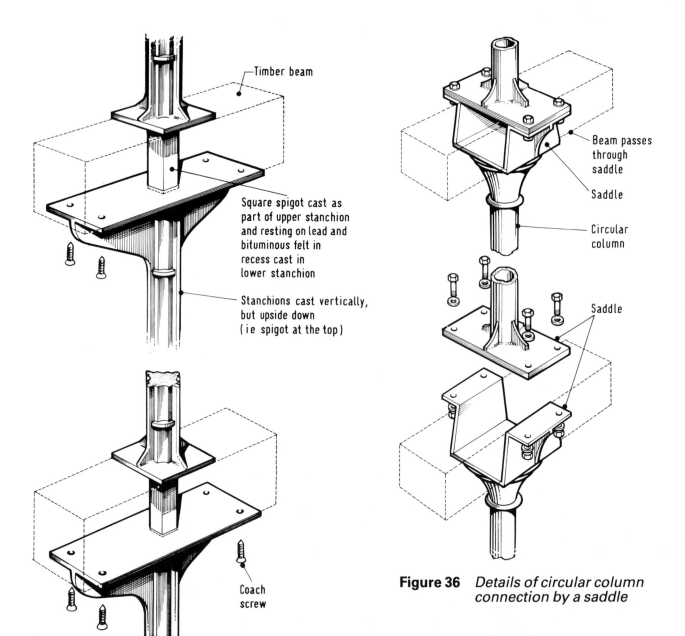

Figure 35 *Examples of cruciform stanchion with flat ends*

Figure 36 *Details of circular column connection by a saddle*

metal, were supposed to heat the joint sufficiently for the resultant casting to be monolithic. However, the process was always dubious for structural work, and became totally discredited after the Tay bridge disaster in 1879, where lugs for securing bracing members were suspected of having been 'burnt-on'.

5.4 CHARACTERISTICS OF WROUGHT IRON

5.4.1 Material structure

Wrought iron is a laminar or fibrous material, being almost pure iron, but with residual impurities in the form of slag which remain to give the layered character. It is anisotropic, and the slag is the limiting factor on the strength of the iron, in tension and compression. The properties are improved by rolling, which tends to draw the laminar slag layers into fibres.

Repeated rolling further improves the properties up to a maximum of about four passes. The number of passes used to be one method of describing the quality of wrought iron, but text books of the period did not appear to distinguish between these grades.

The presence of the slag was not entirely a disadvantage. It was capable of acting as a flux, and it enabled pieces of very hot (but still solid) iron to be joined simply by hammering or rolling. Thus larger wrought-iron sections could be formed by building up and rolling plates together. However, the economics of the time seemed to have favoured rivetting rather than rolling as a means of fabricating girders.

5.4.2 Ultimate strength

The material structure of wrought iron means that its ultimate strength is more variable than that of a homogenous material. Its compressive strength is slightly less than its tensile strength, unlike cast iron. This results from its tendency to delaminate in compression, along lines of slag in the member.

Twelvetrees [24] quoted the following characteristics for wrought iron:

	tonf/in^2	N/mm^2
Tensile and flexural tensile strength	18 to 24	280 to 370
Compressive strength	16 to 20	245 to 310
Shearing strength	75 % of tensile strength	
Elastic limit	10 to 13	155 to 200
Young's modulus (tension and compression)	10 000 to 14 280	155 000 to 221 000

5.4.3 Fabrication and connections

The ability of wrought iron to resist greater tension, and the limitations on the maximum weight of any single component, combined to make mechanical connections in wrought iron much more important than in cast iron.

The process of heating and hammering together sections of iron was known as welding, but it has little to do with the present-day welding process, and it was generally discounted for structural connections. Wrought-iron structural components which could not be formed by heating and hammering pieces of iron were formed by rivetting.

Rivetting was also used to connect wrought-iron members, but the vast majority of rivets used were to connect plates, angles and tee sections to form built-up stanchions, girders and trusses. In its heyday, this process was highly industrialised, and it is therefore not surprising that the design and analysis of such connections was rather different from that used in present-day steelwork.

The principal difference was that the strength of a rivetted connection was considered in terms of a percentage of the strength of the solid metal which it achieved, rather than attempting to vary the capacity to match the theoretical strength requirement, as in modern design.

The result of this is that the pattern of rivetting does not vary according to the stress variations in the member. This, in turn, means that the factor of safety on rivetted connections is likely to vary considerably, and it is therefore important to assess with particular care, those areas which are most heavily loaded, where the factor of safety is likely to be at its lowest. The permissible shear strength of rivets is likely to be similar to that of the parent wrought iron.

5.5   CHARACTERISTICS OF EARLY MILD STEEL

5.5.1 Mild steel

Early mild steel came into common use in buildings in about 1890. Some supplies were obtained at this time from overseas countries such as Belgium and France. Section sizes were first published in 1887 in the Dorman Long handbook, and shortly after in the Redpath Brown handbook. Member sizes and their source were often indicated on the web. Section sizes were standardised by BS 4 [25] in 1903 and later modified in 1921 and 1932.

The permissible stresses in the 1909 London Building Acts were generally used for steel design until the introduction of the code of practice, BS 449 [26], in 1932. Mild steel was significantly stronger than wrought iron and less variable. However, it had a greater tendency to corrode, and it lacked the ability to be 'welded' to itself by the application of heat or hammering. Early welds were most likely to have been carried out by gas-fusion processes, rather than the modern electric arc method. Weldability by modern methods should be checked (see Section 5.8).

Mild steel is commonly found in joists, flat plates or rods, and, like wrought iron, in rivetted plate girders or trusses. The quickest and most reliable means of identification is a metallurgical test. Steel has much less carbon and other impurities than cast iron, but it has considerably more than wrought iron. The first standard specification for steel was BS 15 [27] in 1906 (now BS 4360).

Joists generally had tapered flanges, unlike modern parallel flange beams. A comprehensive description of old joist sizes is available in a recent BCSA publication [22]. The strength of modern steel is based on its yield strength rather than its ultimate tensile strength, which was specified as 28 to 32 tonf/in^2 (450 to 485 N/mm^2). A factor of safety of 4 was usually applied to this ultimate strength to obtain permissible stresses. These are lower than would be used in current design.

Steel rivets were used to provide a non-slip connection. Their permissible shear strength was specified in BS 449 [26] as 5 tonf/in^2 (78 N/mm^2) when installed on site. The rivet diameter is typically 60 % of the diameter of the cup head.

5.5.2 Concrete and 'patent' floors

The scope of this Report excludes modern reinforced-concrete framed structures, but it is quite likely that concrete will be encountered in buildings of the period covered. Such concrete elements may be original or examples of later additions.

Design in reinforced concrete was first covered by the 1915 'Reinforced concrete regulations' to the London Building Acts [6], and the first fully-framed buildings in the UK date from around 1920. However, reinforced concrete was used in floors and internal columns enclosed by load-bearing brick walls in buildings from around 1900.

Many buildings incorporating reinforced concrete were designed to the 'Hennebique' system. L.G. Mouchel & Partners were the consulting engineers who first used this system in the UK, and they maintain record drawings relating to their work of 80 years ago. There are also numerous text books of that time which describe the various systems and methods of analysis. Elastic theory was generally used to proportion the reinforcement.

There were also numerous other patent floor systems, some using wrought iron or steel plates acting compositely with concrete. Some of these systems, which were common in the USA around 1900, have been used in the UK. Many systems were produced on a 'design and build' basis.

As with all reinforced concrete, an assessment of strength depends on the reinforcement which it contains. If the details of the system used cannot be identified, it is extremely difficult to assess its capacity without major works to cut through the section.

In general, the analysis and details of this period lead to designs with factors of safety in respect of shear and anchorage lower than those in bending. Compliance with the then current design methods may not be a guarantee of structural adequacy. In such cases, a load test may be a more effective method.

5.6   DESIGN METHODS AND REGULATIONS

There were no statutory regulations governing the use of iron in building structures until 1909, when the London County Council General Powers Act [6] came into force. The part of this act which applied to iron members laid down minimum metal thickness and permissible working stresses. It also gave tables showing how these were to be modified for 'pillars' to take account of their slenderness and end fixity. Other sections dealt with other materials, and laid down design live loads for different classes of buildings (see Section 2.4, page 17).

The use of permissible working stresses was a marked departure from previous practice, which was based on experience, load testing, or simple analysis of the ultimate strength of members, with factors of safety applied by the designer. The permissible stresses in the 1909 Act [6] (repeated in the 1930 Act [7]) were as shown in Table 3.

Table 3 *Permissible stresses from the London Building Acts*

Material	Working stresses (tonf/in.^{2})				Working stresses (N/mm^2)			
	Tension	Compression	Shearing	Bearing	Tension	Compression	Shearing	Bearing
Cast iron	1.5	8	1.5	10	24	125	24	156
Wrought iron	5	5	4	7	78	78	62	109
Mild Steel	7.5	7.5	5.5	11	117	117	86	171

These are similar to the ultimate strength values for the materials quoted in Sections 5.3 and 5.4, divided by the appropriate factor of safety.

Just prior to the Act, Twelvetrees[24] had recommended the following safety factors shown in Table 4 to be applied to the calculated ultimate strength of a member.

Table 4 *Factors of safety on cast iron, wrought iron and steel*

	Dead load	Live load
Cast iron		
Beams	5 to 6	8 to 9
Columns	5 to 7	8 to 10
Wrought iron and steel		
Beams	3 to 4	5 to 6
Columns	4 to 5	6 to 7

In deriving these values, Twelvetrees discounted 'dishonest manipulation of test results' by the iron master, but listed the following contingencies as having been considered deriving the factor of safety:

1. possibility of errors in estimating the loads
2. differences between the materials in bulk and in the tested specimen
3. possible errors in the calculations of stresses and strains
4. possibility of unexpected strains as a result of indifferent workmanship
5. risk of deterioration from various causes
6. establishment of oscillating stresses
7. fatigue from vibration or shock.

In addition, Twelvetrees required the provision of 'a sufficient margin of safety'.

It can be seen that the LCC permissible stresses [6] broadly relate to Twelvetrees' average ultimate strength values, divided by his dead load safety factors (i.e. 4 for wrought iron and steel, or 5.5 for cast iron). However, the design live loadings required by the 1909 Act were generous, and it is possible that an extra margin of safety was incorporated here. In general, therefore, the application of the 1909 Act was unlikely to give results greatly different from those obtained by previous methods. Paradoxically, the use of cast and wrought iron was in rapid decline by this date.

However, the origin of the stresses laid down in the 1909 Act remains unclear, and there seems to have been no clear contemporary explanation either. *The Builder* (7 August 1909) says 'Section 22 (of the Bye-laws) contains tables for the working stresses in cast-iron and mild-steel pillars, the values being perfectly safe, although not appearing to have been computed by any formula with which we are acquainted'. Nevertheless, the values in the 1909 and 1930 Acts are a good basis on which to begin an appraisal of early structures which used metal.

5.7 ASSESSMENT FOR RE-USE

5.7.1 General

The principles of structural assessment are described in Section 2.3. When an existing iron structure, or a part of it, is being considered for re-use, it is important that the assessment is not based on attempts to adapt modern methods of steel design to the obsolete materials.

It is fundamental to any assessment to first establish the 'quality' of the original construction. Such an assessment does not depend entirely on metallurgical or strength testing, but rather on determining the age of the building, and observing how its overall construction compares with contemporary knowledge.

The proposed use of the building probably takes one of three forms:

1. continued use with loadings and structural form basically unchanged
2. new use with local structural alterations. Over-all loadings unchanged, but adjustments between dead and live loads
3. new use with major structural alterations, major changes (usually increases) in loads and stresses.

The successful re-use of existing iron structures depends on the original structure being of reasonable quality, and its having survived reasonably free from major alterations.

Successful re-use with major alterations is unlikely, even with ironwork of exceptional quality. This is because it is probably fairly well matched to the original requirements. Even if spare capacity can be demonstrated in the ironwork, this should be considered in the context of the capacity of the walls and foundations which support it.

5.7.2   Pillar design

The understanding of strut design has been quite a recent achievement, and 19th Century buildings were built relying on test results rather than on theory. Even the 1909 Act [6] deals with the problem in an unfamiliar way, because it requires the slenderness to be computed as the actual length divided by the least radius of gyration. The designer then identifies the permissible stress according to whether the ends are 'hinged' or 'fixed'. In other contemporary sources, the terms 'pinned' and 'flat' are used. It is clear that these terms were used rather differently than would be the case today. In particular, it seems that the pinned end condition was envisaged only when particular measures (e.g. machining to a radius or the insertion of a genuine pin) were present. This is in contrast to present-day steelwork design, where columns are regarded as pinned unless particular measures have been taken to make them not so. When assessing pillars to the 1909 Act, it is generally appropriate to take a degree of flatness or fixity at the ends.

Figure 37 presents the permissible working stresses for cast-iron pillars with various forms of end restraint. The columns in Figures 34 to 36 (pages 57 and 58) are normally regarded as 'fixed' or 'one-end hinged' and that in Figure 38 as 'pinned' end. Also shown for comparison on Figure 37 is the corresponding line for a mild-steel pillar with pinned ends.

Although most buildings with cast iron were built before the 1909 Act, Figure 37 is generally adopted by building control officers to assess pillars for their current use. These stresses probably take into account some eccentricity and moment transferred from the beams, because the permissible stresses for 'short' columns are well below those of the basic material (Table 3).

5.8   SAMPLING AND TESTING OF IRON AND STEEL

Detailed testing of iron structures to justify higher strengths than in Table 3 is rarely practical, because the variability of cost and wrought iron is such that a relatively large number of test samples are needed to establish the strength of the material. The average strength of the samples should be considerably higher than the design ultimate strength (or 95 % confidence limit), because the scatter of results would be considerable. The samples should also be taken from representative parts of the structure. However, the specimen size is normally quite large compared to the member size, and samples could not be taken from very highly stressed parts of the member. If strength tests are carried out, the following approach may be used as a guide to the statistical principles:

For 6 to 10 samples, the 95 % confidence level is in theory the average strength minus two standard deviations (SD) of the test results. For two samples, this is the average minus 6.3 SD, and for a very large number of samples the average minus 1.65 SD. Therefore, around six samples from a series of similar elements usually suffice, provided that the material chosen is reasonably representative of the quality of the remaining structure, including its likely defects. This latter factor is often the most difficult to resolve.

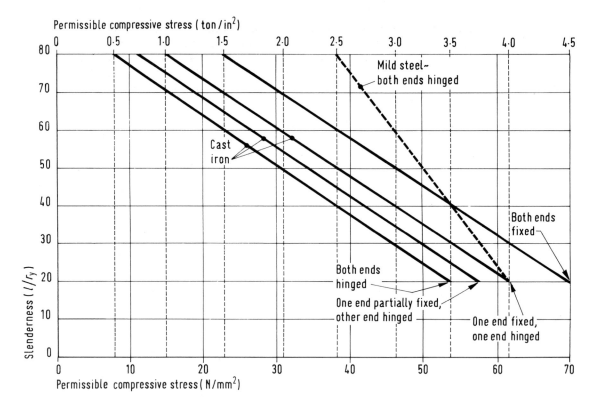

Figure 37 *Permissible stresses in cast-iron and mild-steel columns*

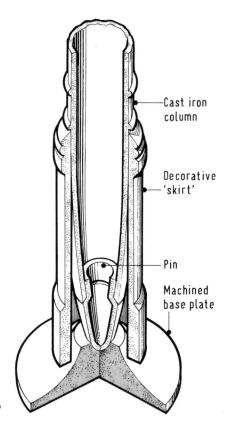

Figure 38
Example of pinned column base

Cast iron is sensitive to the size of the section and its position within the casting. If tests are taken of the thickest material, lower strengths result when compared to thinner material. Another consideration is its porosity and casting defects. In view of this, a reduction factor of 3 SD (for six samples) may be more appropriate in determining its design ultimate strength. However, permissible stresses much in excess of those in Table 3 are not normally accepted by the Building Control Officer.

The standard tensile test to BS EN10 002-1 [27] requires that the samples should be fairly large, so that up to 100 x 200 mm specimens may be needed, allowing for the removal of the zone near the cut. In some cases, a 50 x 100 mm wide specimen may suffice. Removal of large samples in cast iron may be impractical.

In addition to laboratory tests, it is often useful to carry out non-destructive tests (e.g. hardness surveys), using a portable Brinell hardness tester, which correlates roughly with the ultimate tensile strength.

The thickness of hollow pillars can be determined by drilling three equi-spaced holes on the circumference to account for any irregularities in the shape of the cone. Ultrasonic techniques are rarely successful.

When carrying out a metallurgical or chemical test, a relatively small sample and even drill-swarf may be used. This accurately defines the constituents of the metal, and it should be carried out by a specialist firm. The weldability of early mild steel is related to its carbon content and other residual elements, higher levels making it more brittle and less easy to weld. It is not possible to weld cast and wrought iron for this reason using modern welding methods. Later mild steel can be welded.

5.9   IRON IN FIRE

The current approach to the provision of structural fire protection has developed since iron went out of general use as a structural material. There is therefore no substantial body of test information on the behaviour of iron, whether protected or not, to BS 476: Parts 20 and 21 [20].

The adaptation of iron structures to new uses, with the minimum change to their appearance, has led to a re-examination of their performance. The paper by Barnfield and Porter [29] describes recent work on cast iron in some detail.

Broadly speaking, the performance of cast iron in fires is better than one might expect, based on the experience of the performance of steel structures in fires. It is possible for hot iron to crack under a stream of cold water, but cases where this has led to premature structural collapse are rare. Records of real fires involving cast-iron structures are currently being examined to study the effects of these factors.

Barnfield and Porter show that cast-iron members generally heat up slower than equivalent steel members, because of their lower perimeter/area ratio (P/A). The strength of cast iron also decreases less rapidly with temperature. Typically, its strength reduces by 50 % at about 650 C, compared to 550 C in steel. Cast-iron members are also stressed to lower levels than steel, and the ways in which they were incorporated into buildings usually allow them to expand without generating increased stresses. When cast-iron columns support timber beams, the beams may char, but they do not expand and give rise to lateral forces and movements on columns as might steel or iron beams. This expansion is one explanation of why some cast-iron structures behave worse than others in fire.

Thus there is every reason to consider individual cases of cast iron in fire on their own merits, rather than to apply general rules. When this is done, capacities of ½ h in (unprotected) cast-iron structures are common, and periods of up to 1 h can be achieved.

The use of the P/A factor as a basis for predicting the temperature rise in a member is applicable to beams as well as to columns. In cases where the beams are partly embedded in brickwork or concrete (e.g. in jack-arch and filler-joist floors), greatly enhanced performance results from this partial protection. A full description of the P/A method is given in Reference 30.

Where the structural fire resistance of a cast-iron member is to be increased, it is no longer necessary to apply protective materials which destroy the shape and character of the member. Several different intumescent materials are available, which can be sprayed onto the iron after suitable cleaning and priming. There are 'thin-film' and 'thick-film' intumescents, the former being usually described as intumescent paint and the latter intumescent mastics.

Both work on the principle of swelling up at high temperatures to provide a charred 'foam' which so insulates and protects the member from heat that the temperature of the member remains well below that of the fire. Both intumescent materials are applied by spraying. Thin-film materials require no further treatment, but the thicker mastics (typical thickness 5 to 10 mm) require to be manipulated when wet to reinforce sharp corners and to achieve an even finish.

However, the maintenance of structures with intumescent coatings should be carefully considered, particularly where the covering may be removed during the life of the building. Such measures should be discussed with the Building Control Officer early in the scheme.

6. Foundation investigation and strengthening

Many old buildings do not have true foundations (i.e. a well-defined and proper interface between the superstructure and the ground), but they continue to resist the loads from the building without apparent distress. Nowadays, even the most modest of new buildings on good ground tend to have mass concrete strips beneath the walls to give relatively low bearing pressures.

It does not follow that the means of founding older, and often more flexible, buildings were inadequate for their original use. Indeed, this compatibility between the structure and the ground has developed over the relatively slow period of construction and the long life of the building. Thus the net foundation bearing pressure which one might assign to the ground in designing a modern structure is not the sole criterion for assessing adequacy.

The essential steps in considering the foundation of a building to be renovated are therefore:

1. identify the form of superstructure and foundations as they exist
2. determine whether they are 'compatible' at present (i.e. the extent of any prior movement and cracking of the structure)
3. determine how the proposed alterations might affect this compatibility (e.g. increased load or movement, undermining, etc.)
4. devise ways of eliminating or reducing these effects without introducing new problems (e.g. stiff points).

There are various means of upgrading foundations, depending on the scale of the structure, which are described in Section 6.2.

6.1 INVESTIGATION OF FOUNDATIONS

The first stage of any assessment of existing foundations involves identifying the over-all form of the superstructure, substructure, and the soil, then determining how well they act in combination. Consideration of one without the other is likely to lead to solutions which may aggravate rather than relieve the situation.

The superstructure can usually be easily examined, but trial excavations are generally needed to identify the form of foundation. The investigation should take the following stages, based on BS 5930 [5], and described in CIRIA Special Publication 25 [31]:

1. visual inspection of the building and its neighbours, noting any cracking and distortion (see Section 2.2)
2. examination of old drawings, local plans, and those of the statutory authorities. A check list of sources is given in BS 5930
3. examination of the geology, topography, and water level records of the area, including nearby trees, old water courses, etc.
4. exposure of the substructure and foundations, often by trial pits, supplemented in some cases by boreholes, to assess the general ground condition.

A knowledge of old methods of foundation construction is useful to anticipate what may be found. Examples might be heavy brick corbels, inverted vaulting or timber piles, and capping beams. Alterations may have taken place, in which case the twin risks are that a local modification may be found and taken as being typical, or that a heavily-modified area may be missed. Consider the example of Figure 39, showing that if three pits are dug at different locations along the length of a wall, the general foundation type may be interpreted in different ways, and the bearing capacity may be seriously over or under estimated.

The ground investigation is not considered to be 'completed' prior to construction, and re-assessment may be necessary at a later stage. The scale of investigation should be appropriate to the scale of structure, and the consequences of any error considered in the interpretation of the results.

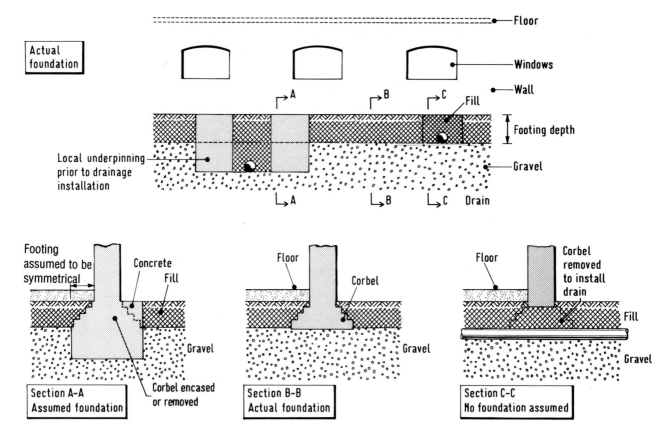

Figure 39 *Examples of misinterpretation of single trial pit information*

Examples of those factors which may lead to incompatibility between the foundations and the superstructure, and therefore may require some form of upgrading of the foundation, are:

1. undermining of existing foundations such as by new drainage or loss of lateral restraint to the foundations by nearby excavations, contraction of basements, or landslip, etc. (Figure 40)

2. changes in the moisture content, leading to movement of shallow footings in clay. This can be the result of adding or removing trees or changes in the groundwater level or pattern of water run-off from the building and its surroundings

3. destroying or altering the stiffening effect of the walls or the load distribution pattern on the foundations (e.g. by forming new openings). It is relative movement, resulting in shear or tensile strains, which causes cracking in walls

4. deterioration of steel, concrete or timber in the foundations by chemical attack, insect attack, or groundwater level changes

5. addition of extra loads which may make high differential settlements likely

6. addition of extra loads which may cause failure of piles (particularly timber piles)

7. introducing new sections of brittle construction which cannot accommodate continued movement of the foundations

8. introducing stiff points in the foundations which cause increased relative movement between the more flexible and stiff parts

9. removal of overburden, leading to ground heave and reduction in bearing capacity

10. vibration effects from nearby traffic or change of building use.

Settlement is a time-dependent process, and if the building is sufficiently ductile to accommodate differential settlement over a long period, it is often reasonable to apply further loads which would have been unacceptable had all the loads been applied together at the outset. During the original construction period of 2 to 3 years, all the movement in gravel or sandy soils, and a considerable proportion of that of clay soils, occurs. The period of consolidation of clay soils can be considerable, and settlement or heave can also occur as a result of changes in the moisture content of the soil. Foundations on sandstone, limestone or other rocks do not settle appreciably, but weathered chalk can behave like clay.

The requirements of BS 5930 [5] may be used as a guide to the techniques of site investigation, although the Code is not directly applicable to renovation work. Trial pits should be made close to, and down to, the existing foundation level, preferably both just outside and within the building. Deep excavations may undermine the foundations, and excavations near windows or piers should be avoided. Undisturbed and natural moisture content samples of clay in cohesive soils, and disturbed samples of uncohesive soils with *in-situ* density measurement, may be taken by a specialist. Simple vane tests may be used to obtain an *in-situ* measurement of the strength of clays. Photographs are a useful record of the soil strata and foundation to supplement cross-sectional drawings and measurements.

A reasonable number of trial pits should be dug, but boreholes may be preferred to obtain data on the soil conditions at depth (e.g. where piles are to be used). Boreholes may also be advantageous where the watertable is near the ground surface. In straightforward areas without height restriction, normal boring and testing equipment such as the light cable percussive rig (the so-called 'shell and auger') may be used. In cases of low 'headroom' or difficult access, small rigs which can be man-handled down stairways are necessary. In suitable soils, such as clays, the hand auger may be used.

To obtain some measure of the bearing pressure, the Standard Penetration Test (Test 19 in Reference 32) should be used, recording the blow count for each 75 mm of penetration and totalled over 300 mm. This test needs headroom to allow the dropping of a 65-kg weight through 760 mm. It also needs room to fix on the tools and rods. The non-standard lightweight Macintosh probe type tools may also be used, and the penetration results may be calibrated against nearby SPT results.

Information on the watertable and its seasonal movement is very important, as are water samples for chemical checks. Examples of unsuitable ground materials which may give rise to continued movements are unconsolidated fill materials, peat strata, soft organic clays, and expanding shales.

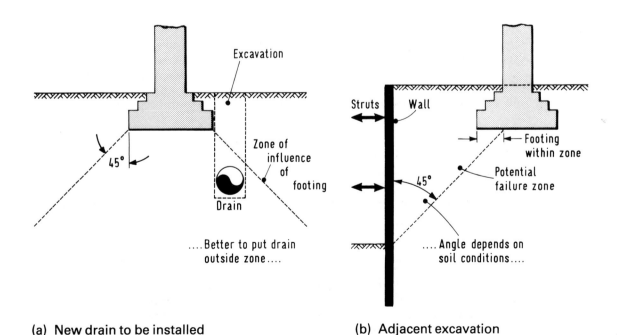

(a) New drain to be installed   (b) Adjacent excavation

Figure 40   *Cases where underpinning of footing may be necessary*

6.2 UPGRADING OF FOUNDATIONS

6.2.1 Background

The simplest means of avoiding the creation of additional forces or 'incompatibilities' between the structure and existing foundation is to devise an alternative scheme which eliminates the problem. Examples are re-routing of new drains, or construction of independent foundations to support any new superstructure. In some cases, it may be feasible to replace upper walls by lightweight blockwork if additional loads or new storeys are to be introduced.

Where new foundations are designed to act independently of the old, some provision for relative movement should be made. This is most often seen when an extension is built onto the side of an existing building (Figure 41).

If it is concluded that upgrading of the foundations is necessary, a number of techniques are available. The choice rests mainly with the desire to avoid creating new problems during the remedial work, and the efficiency of the site operation. The main options in foundation upgrading are widening to reduce bearing pressures, and deepening to found on stronger ground or ground less affected by the construction work, loads from adjacent buildings or seasonal movements.

Many of the techniques can be carried out only by 'specialists', and they therefore have implications on the programme of work, access etc. Figure 42 illustrates some of the methods of upgrading foundations. Some are traditional methods based on continuous or partial underpinning, others are proprietary piled systems. The decision as to which system to use depends on:

1. the nature of the ground, depth of suitable strata, the watertable, and the existing foundation
2. the risk of damage to the structure during the work
3. whether the work is part of major refurbishment or remedial work to an occupied building
4. the nature of the adjoining structures, and the effects of the work (e.g. noise, vibration) on them
5. effect of any 'stiff points' created on the performance of the building (e.g. differential settlement)
6. 'total cost' of the operation, including that of the 'specialist', opening up, access, downtime, and reinstatement.

6.2.2 Conventional underpinning of walls

The physical operations involved in continuous underpinning of foundations to walls are fairly well known, but there are particular aspects which should be emphasised. Underpinning may be used to form deeper or wider foundations, but it relies on adjacent excavations which need to be shored and extensive enough to allow access for concreting.

Underpinning is described in standard textbooks and BS 8004 [33, 34]. Each excavation should expose only 1 to 1.4 m length of wall, which should be capable of arching over the excavation. If the quality of the brickwork is suspect, the excavation width may need to be reduced (to around 0.7 m). Attention should be given to the position of openings, piers and cross-beams, so that sections of the structure carrying heavy loads are not left unsupported. Underpinning of these elements is not recommended without a special support system.

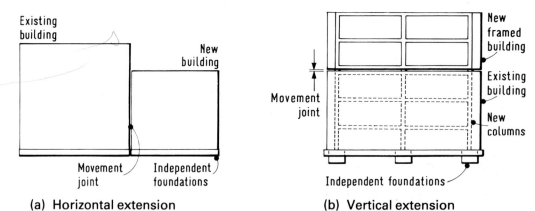

(a) Horizontal extension   (b) Vertical extension

Figure 41 *Provision for relative movement between new and existing construction*

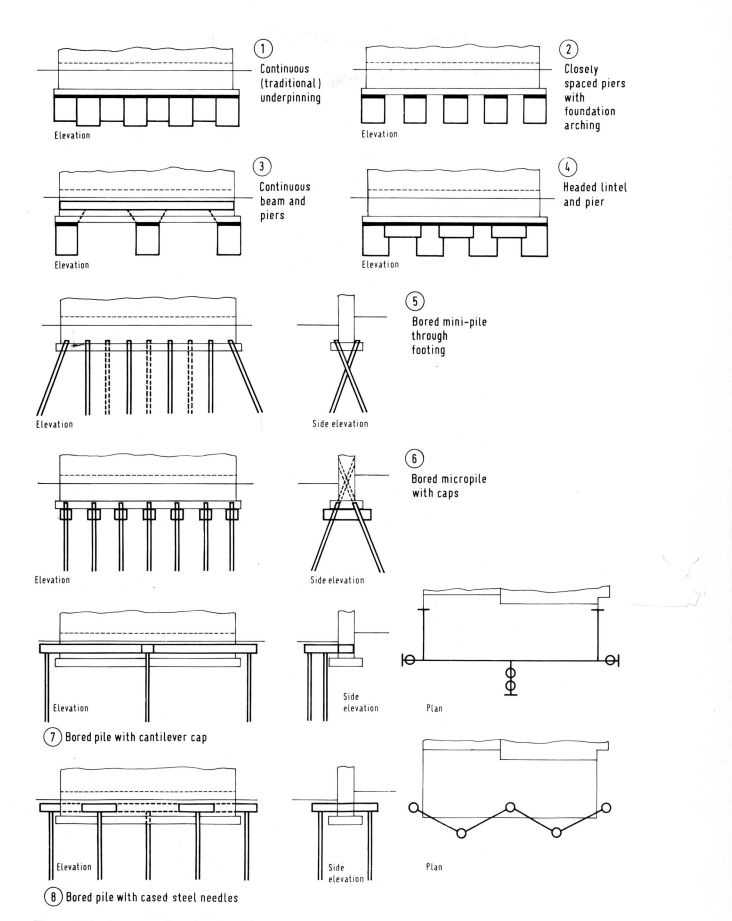

Figure 42 *Forms of underpinning*

1. Continuous (traditional) underpinning — Elevation
2. Closely spaced piers with foundation arching — Elevation
3. Continuous beam and piers — Elevation
4. Headed lintel and pier — Elevation
5. Bored mini-pile through footing — Elevation / Side elevation
6. Bored micropile with caps — Elevation / Side elevation
7. Bored pile with cantilever cap — Elevation / Side elevation / Plan
8. Bored pile with cased steel needles — Elevation / Side elevation / Plan

The progression of underpinning along the wall is normally in groups of five short lengths which are excavated in the order 3 1 4 2 5 3, etc., where the first excavation is not at a sensitive part of the building (Figure 43). The trench is then dug and shored to the appropriate depth beneath the existing foundation, and the underside of the footing is dug away and cleaned. Collapse of the soil behind the footing should be prevented by shoring at this stage. When the bearing stratum is reached, the concrete base should be poured to within about 75 mm of the underside of the footing. In most cases, low-grade structural concrete is satisfactory, provided it has adequate sulphate resistance. If the concrete is poured in two levels, it should be keyed in with the preceding pour. When hardened, the gap between the concrete and the footing is then packed with 'dry-pack' concrete. This concrete normally consists of 3 to 1 proportions of sand to cement mixed to a dry consistency, and it should be well rammed in to limit further movement. In traditional underpinning, slates were often used as wedges.

Where the ground quality does not increase significantly with depth, or where there are groundwater problems, widening rather than deepening the foundation may be more practical. Deep excavations need considerable shoring (Figure 44). Any temporary shoring left permanently in place should not decay (which would leave voids in the ground).

Partial underpinning of a wall may create further problems, because masonry may crack if there is continued relative movement, particularly in the case of hogging movements at 'stiff points' (see Figure 13, page 33). For this reason, partial underpinning may not be the best solution.

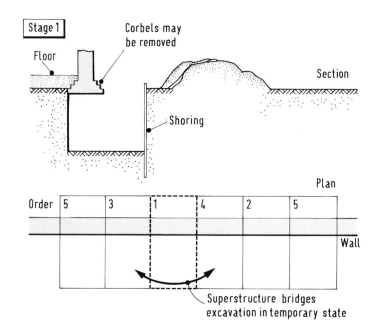

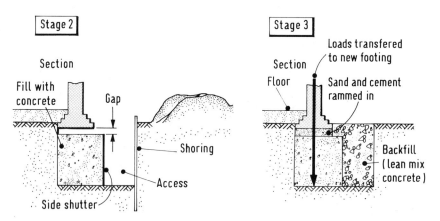

Figure 43 *Sequence of underpinning of a wall*

6.2.3 Other underpinning techniques

There are various techniques for temporarily supporting structures to permit modification or strengthening with minimum disturbance. The technique of 'needling' or use of dead shores has been mentioned (Figure 45). A hole is cut through the wall and a steel cross-beam or 'needle' inserted so that the wall thrust is transmitted to supports and temporary foundations which are well clear of the excavation to be made. This depends on access and adequate space on both sides of the wall, although 'cantilever' methods can be used (see Figure 42, page 71). To support a column or pier, it is necessary to supply a collar or bracket or similar arrangement to connect the column to the needle beam or pair of beams. Underpinning of elements subject to high load requires careful design and planning.

When the underpinning work is complete, it may be necessary to introduce a jack system or wedges to compress the ground and to ensure load transfer prior to removal of the needle beam (see Section 6.2.5).

Underpinning may be necessary if adjacent excavations cause reduction in lateral restraint to the ground beneath the foundation. This is shown in Figure 40 (page 69), where as a guide the horizontal zone of influence of the excavation is taken to be equal to its depth. Therefore, excavation within a 45° projection from any loaded area, as would be the case for the installation of the drain beneath the footing in this example, is not recommended [32].

If the excavation is properly strutted and expected horizontal movements are small, this 45° projection is probably conservative. However, any decision depends on the sensitivity of the building to movement and vibration, both during and after the excavation work.

One proprietary 'beam and stool' method has been developed which minimises disturbance to the structure (Figure 46). Short sections of the wall are cut away, and the stools are inserted and packed. The remaining sections of wall between are then cut away, and reinforcement is threaded between the stools. Following this, the whole arrangement is concreted. When this ring beam is completed, the building is then effectively supported on stiffened strips.

Another modification to this system is the headed pier and lintel form, where deep underpinning piers are constructed and lintels inserted between the second stage excavation (Figure 42, page 17). This is advantageous where the bearing stratum is very deep.

Where it is proposed to widen, rather than deepen, a pad foundation or footing without underpinning, it is often not possible to connect the old and new parts of the footing adequately in order to resist the applied shear and bending moment across the footing, except where reinforcement or post-tensioning strand or rods can be inserted. It is also necessary to ensure that the ground beneath the new part of the footing is properly compacted. For these reasons, conventional underpinning to create a wider base is often preferred.

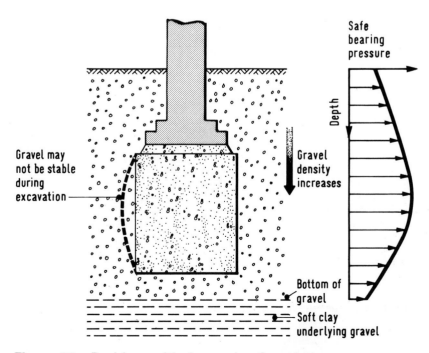

Figure 44 *Problem with deepening foundation*

6.2.4 Use of mini-piling systems

The decision as to whether to pile or to use one of the underpinning techniques previously described depends on cost, access, depth of foundation and groundwater level, obstructions, noise, vibration and extraction of the debris and soil. The simplest system is to form short bored piles using a hand- or machine-operated, small-depth auger. The holes are then filled with concrete and reinforcement, which may be tied into a ground beam constructed to support the wall or column at a later stage.

It is also possible to make a direct load transfer to the structure by penetrating the existing foundation and providing a shear key between the pile and the foundation (stitch piling). The various methods of piling, as described in Reference 35, are:

1. percussive driving – the systems include top drive (using a mole or hammer), bottom drive, or pre-drive, followed by construction of pile. In its simplest form, the drive casing is left in, and it may be grouted up

2. usually pre-cast units pushed against the structure by a hydraulic jack

Figure 45   *Examples of needle beams to temporarily support wall*

(a) Holes cut for stools which are inserted and packed up

(b) Adjacent wall section cut away and reinforcement inserted

(c) Whole section of beam concreted and packed up

Figure 46 *Proprietary stool system for underpinning*

3. drilling – using rotary cutter, auger, or rotary percussive methods, with or without temporary casing
4. boring – percussive cutter and shell, hand auger or chisel in rock and boulder.

Pile diameters used for underpinning range from 50 mm for micro-piles as in Figure 47 to 300 mm for bored piles with a percussive or cable-boring rig. The usual diameters of drilled and driven piles are 75 to 150 mm. Rotary methods have the advantage that vibration is reduced, and mud circulation may be used to stabilise the hole. Rotary or rotary percussive methods are best for drilling through the existing structure. Low-headroom bored pile rigs with shortened legs are often used in renovation work (Figure 48a).

Examples of underpinning, including the use of mini-piles, are shown in Figure 42 (page 71). Figure 49 shows the underpinning of a façade and new support beams and columns by mobile rigs forming raking piles. The system of a network of piles (or pali-radice) has been used for ground stabilisation and foundation strengthening.

The alternative to 'stitch piling' by drilling through the existing foundation is to pile symmetrically on either side of the wall, and to install a pile cap or cross-beam by underpinning beneath the wall. Cantilever pile caps can be used where it is not possible to pile within the building, but this has the disadvantage of increasing the load on inner piles.

Mini-piling systems are particularly attractive when the ground conditions otherwise dictate the use of deep excavations, often in the presence of groundwater. Typical rotary piling rigs are shown in Figures 50 and 51. Because of their small diameter, the piles are normally designed for shaft friction, not end-bearing strength. Working load capacities range from 20 to 30 kN for the small diameter piles (50 to 75 mm) up to 1000 kN for the larger diameter piles (200 to 400 mm). The smaller capacity piles are generally spaced close together along a wall, and where 'stitch piles' are drilled through the existing foundation it is often the quality of the foundation which determines the pile capacity. A pile test can be invaluable where the design parameters are ill defined.

text continues on page 79

Figure 47 *Use of small-diameter powered auger*

Figure 48b *Single bar reinforcement in small diameter piles*

Figure 48a *Low-headroom pile rig being used to suspend reinforcement cage during fixing in small diameter bored-pile*

Figure 49 *Rotary-driven piling system used for underpinning large masonry wall*

Figure 50 *Rotary-driven piling rig in operation*

Figure 51 *Piling rig used in confined space*

Raked or inclined piles usually have to resist bending moments. In many small-diameter piles, only a single central reinforcing bar is used (Figure 48b), and it may be considered desirable to extend this reinforcement down the pile. Circular hollow steel sections may be used to improve the ability of the pile to resist bending. For permanent works the tubes should have sufficient protection against corrosion. Where balanced raked groups of mini-piles are designed for axial loading only, the angle to the vertical should be limited to 15 to 20°.

Many forms of drilling involve considerable use of water to remove the soil and foundation debris. When selecting the piling system to be used, it is important that it should not lead to excessive loss of ground. Hole stabilisation by casings or 'drilling muds' may need to be considered in difficult alluvial grounds. Some piling methods may be hindered by underground obstructions, and jacked or driven piles may not be possible.

6.2.5 Load transfer

In all forms of underpinning one of the main requirements is for satisfactory load transfer to the existing structure. This is not only to ensure adequate capacity, but also to prevent subsequent movements which may further damage the building. It is relative angular movement which causes cracking. However, all underpinning operations carry the risk of minor movements of the superstructure as it adjusts to its new support.

In 'needling', folding wedges and jacks are sometimes used to develop load and to take up any slackness in the temporary support system. Steel wedges have been developed with precise tapers to control the amount of vertical movement. A typical modern re-usable drawbolt operated wedge has a depth of 80 mm and a movement of 25 mm.

When the new foundation is complete, it is possible to precompress the foundation against the existing structure in order to develop some initial settlement of the ground (and therefore to limit further movement). In cohesive soils, settlement can occur gradually over many months. This compression can be introduced by 'flat jacks', inflated and set to the estimated foundation load. The gap around the jack is then packed up by wedges or steel sections and the jack re-used. Resin filled 'flat jacks' can be left permanently in place where re-use is not feasible.

In 'stitch-piling', direct load transfer is achieved by a grout-shear key between the pile and the existing foundation. In other piling systems requiring a pile cap, a similar method of load transfer to those described above is required.

6.2.6 Other methods of foundation upgrading

Grouting can be a versatile means of underpinning structures and upgrading foundations, by filling voids and compacting loose, cohesionless soils. Grouts may also be used to stabilise the ground prior to an excavation, and to improve the integrity and capacity of existing masonry foundations.

There are a variety of cementitious and chemical grouts which may be used in different cases, depending on the soil conditions, and reference should be made to a specialist. Injection points in the ground are usually of the order of 2 m apart and over 3 m in depth.

A technique which has recently been improved is that of jacking up foundations by pressure grouting, caused by the compaction of a bulb of thick injected grout. This requires careful monitoring of the building movement during the injection process, and it is a specialist operation.

7. Retention of existing façades of buildings

There are many reasons why the architect, developer or local authority wish to preserve the façades of buildings to maintain their outward appearance while the internal structure is partially or completely demolished and replaced. One of the most important legislative constraints is the statutory protection of buildings of architectural or historic importance. Lists of such buildings are compiled by the local authority under Section 54 of the Town and Country Planning Act of 1971 [36]. There are also a much greater number of buildings in conservation areas which, although not listed, cannot be demolished without planning consent. The most common feature to be safeguarded is the external façade, but there may be internal features (e.g. staircases and other fittings) to be preserved.

Façade retention or 'façadism' has become a popular means of satisfying the planning requirements, at the same time providing the client with an internal building structure more suited to the users' needs. It can be applied only to traditional structures with external load-bearing, and inherently robust, walls.

Some façades may be structurally unsound, and collapse during renovation is possible if the existing condition of the façade is not carefully assessed and any prior repair carried out. In Sections 7.1 to 7.6, the common techniques and technical problems in façade retention are examined, drawing upon case examples where appropriate.

7.1   ASSESSMENT OF EXISTING FACADE

The stages in the assessment process are described in Section 2.3. The site survey should reveal the following information about the façades, and their relationship with the rest of the building:

1. The principal structural members. How they fit together, and how they are connected to the façade.
2. The dimensions of the façade, including out-of-plumb and bow.
3. The form of construction of the façade, and the materials used (e.g. facing and backing material).
4. Defects in the façade which can be seen, and likely defects which are hidden.
5. Alterations to the façade during the life of the building (e.g. openings, extra storeys, new facing).
6. The conditions below ground (e.g. basement level and foundation), including any earlier underpinning.

At this point, the proposed structure and the existing façade should be outlined in sketch form. In particular, any changes between the old and new structure should be noted, such as differences in floor levels, or environment. It is then necessary to carefully consider how and where any new connections may be made to the façade, and whether it needs to be repaired or strengthened.

7.2   DESIGN CRITERIA FOR FACADE RETENTION

Load-bearing external walls in traditional buildings resist vertical loads from the floors, and lateral loads from wind. They may also stabilise the internal frame, which may be pin jointed. Similarly, the floors of the building serve to transfer in-plane forces from wind on the façade to walls, cores or braced sections across the building. When the existing structure is partially or totally removed, the façade should be supported temporarily until it is tied into the new structure.

The common technical problems include:

1. designing the temporary support to the façade during demolition and reconstruction of the interior of the building
2. allowing for access of plant and materials, and vehicular or pedestrian movement outside
3. connecting the new structure to the façade, and allowing for subsequent differential movement
4. designing the new foundation system which does not impair the stability of the façade.

In most schemes, the façade is not retained in order to provide a load-bearing function, and normally the new internal frame and bracing system are designed to resist all the applied loads. The retention system may be external or internal to the building or a combination of both. Clearly, external systems offer the main advantage of not interfering with the subsequent reconstruction, but they suffer from the disadvantage of occupying land outside the building line and possibly obstructing foot paths and roads. Internal systems can be designed to cope with the problems of demolition, but they require much more careful planning, and they may be slower. Combined systems can be advantageous. Support for site huts and storage of equipment can be an important factor.

7.3 DESIGN LOADS ON RETENTION SYSTEM

The façade support system has to be both strong enough to resist the lateral forces on the wall and stiff enough so that eccentricities are not created. In general, the requirements of the Code of Practice for falsework [37] should be followed.

The design forces for wind loading should include not only the area of the façade (with a higher pressure coefficient for an individual wall than the completed building) but also the exposed area of the support system, huts, and hoarding. The 1-in-50-year wind speed specified in Codes of Practice [38] is conservative, and BS 5975 [37] gives reduced wind speeds for temporary work (typically 77 % of the above value for a 2-year design life). An additional lateral force to take account of deflections and non-verticality, or alternatively, 2½ % of the vertical load on the façade should also be considered. Impact forces are rather more problematical. Nevertheless, an allowance of, say, 25 kN at the base of the structure, independent of the other lateral forces, should be satisfactory, except for vehicle collision.

The foundation of the support system should generally provide a minimum factor of safety of 1.5 against overturning or 2.0 against sliding. Because of the lack of vertical load on the support system, the weight of concrete or kentledge is normally more critical than the foundation bearing pressure. Where the support foundation and existing foundation to the façade may interact, some form of prior strengthening of the façade foundation may need to be considered.

The members in the retention system are designed in a standard manner, using the load factors or permissible stresses in permanent works. Particular considerations should be given to the joints and means of providing over-all stability by bracing. Slotted or oversize bolt holes may introduce excessive movement, leading to cracking or distress in the façade. The means of attaching the new support to the façade, and then the façade to the new structure, are discussed in Section 7.6.

7.4   CHOICE OF FACADE RETENTION SYSTEM

The choice between timber shores, scaffolding, or a purpose-designed steel system depends mainly on the scale of the work and the space available. The generic types of system are:

1. raking supports external or internal to the building (Figure 52). The main factors to be considered are the density of the support system, the means of access, and the shear and uplift forces applied to the façade. An example of the use of scaffolding inside the building is shown in Figure 53.

2. horizontal trusses or walings spanning internally between vertical trusses which are stabilised by the existing flank walls (Figure 54). In some cases, it may be possible to directly connect the horizontal elements to the flank walls as in Figure 55.

3. braced towers with horizontal trusses or walings spanning between as in Figure 56. These may be used internally or externally as in Figures 57 or 58, and they are advantageous where access or workspace is important. In the example given in Figure 59, a haunched portal frame has been introduced to replace the bracing system at ground level.

4. continuous-braced columns and beams such as in Figure 60.

5. façades supported by internal walls, while the floors are replaced. This can be achieved by walings and internal struts as shown in Figure 61.

6. façades strutted across the building and horizontal forces transferred to flank walls. The in-plane stiffness of the floors is to be replaced by a horizontal bracing system. A variation of the use of mutual support from existing walls is shown in Figure 62.

7. façades supported vertically, while large openings are formed as in Figure 59.

Clearly, there are many variations, and the choice depends principally on the means of demolishing and replacing the existing internal structure. An internal support system has to be installed during a controlled sequence of demolition, because the floors provide the primary means of lateral support to the façade. One scheme would be to install the horizontal trusses on the existing floors as in Figure 54, then to break through the floors to position the support columns. Once the system is assembled and tied into the façade by collars, the remainder of the structure can be demolished.

Ideally, where any internal system is used, the beams and columns in the temporary structure should also form part of the permanent structure. An additional problem is that of forming the foundations to the new structure prior to demolition, because of restricted access and space. Some form of partial demolition may be need to be considered, particularly for the installation of a tower crane.

7.5   CONNECTION OF FACADE TO ITS SUPPORT

A problem in all projects is how to connect the façade to its temporary and permanent means of support. In the temporary stage, a collar is normally inserted on either side of the wall connected to the waling system as in Figure 63(a). This clamps the wall to its support, using folding edges to accommodate unevenness of the surface. The local strength of the wall around these attachments should be checked.

The permanent connection is more difficult, if free relative vertical movement between the façade and the structure is not to be inhibited. One form of fixing to masonry or stone façades is the resin anchor, where a tie rod and a resin cartridge is pushed into a preformed hole. However, the resin anchor should be used with caution where the quality of the surrounding masonry is suspect, and where it may be exposed to fire.

A more positive connection, using a steel plate and rod fixed to the beam by an angle section, is shown in Figure 63(b). The number of fixings should be more than sufficient to account for any variability in the façade material. The angle may use slotted holes to permit any sliding resulting from differential movement, and it may need to be stiffened periodically if it is required to transmit large horizontal forces. Slippage between the wall and the encased steelwork should be permitted by a sliding membrane. The amount of relative vertical movement (e.g. from settlement) depends on the new foundation loads, but as a minimum 5 to 10 mm should be allowed.

7.6   FOUNDATION CONSIDERATIONS

It is often found that foundations to old buildings are weak and susceptible to disturbance. It is therefore common practice to install the foundations for the temporary support and new structure clear of the existing façade. The columns for the new structure may therefore be set inside the building and the floors may be cantilevered to the wall. Some settlement of the façade may still occur, and this should be monitored during construction.

Where the same new foundation is used for the façade and the new columns, their bases should be designed to minimise the eccentricities resulting from their different vertical loads. Underpinning to form these new foundations may cause cracking in the façade, which should be carefully monitored. Means of foundation strengthening are discussed in Section 6.2 (page 70).

The new foundations within the building may use the existing column bases, or they may be formed clear of possible obstructions. Where these foundations are dug or piled prior to demolition, consideration should be given to access and head room.

text continues on page 91

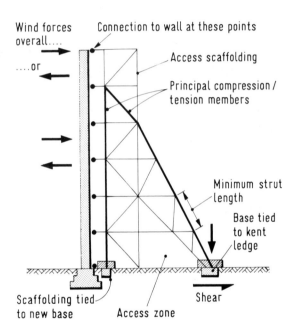

Figure 52
Forces on temporary wall support by scaffolding

Figure 53 *Example of temporary support to wall by scaffolding*

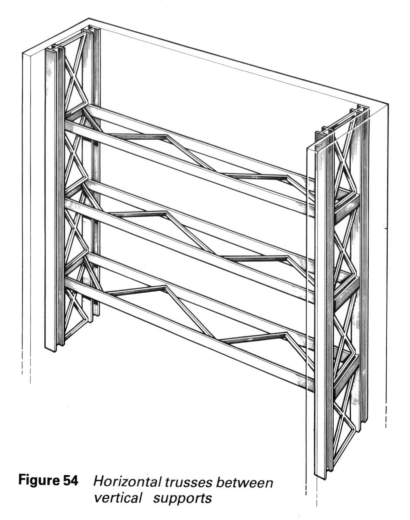

Figure 54 *Horizontal trusses between vertical supports*

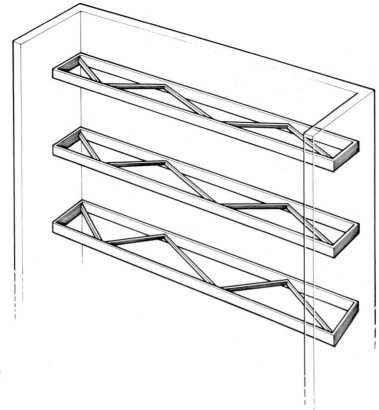

Figure 55
*Horizontal trusses
between flank
walls*

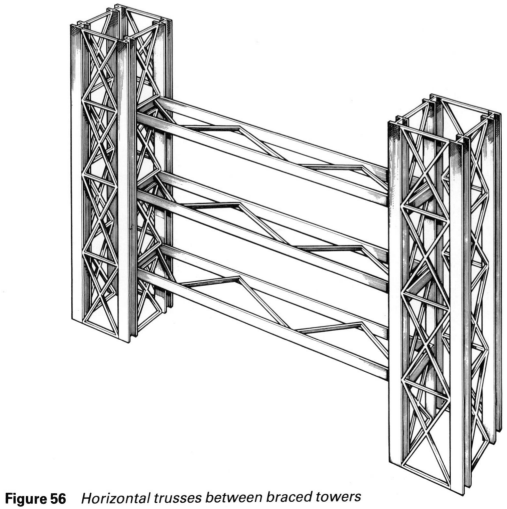

Figure 56   *Horizontal trusses between braced towers*

Figure 57 *Internal support to façade by towers and horizontal trusses*

Figure 58 *External support to façade by towers and horizontal trusses*

Figure 59
*Façade support
system, supported
on portal frame*

Figure 60 *Continuous external support to façade*

Figure 61 *Support to façade of small building by walings and struts*

Figure 62 *Mutual support of existing walls*

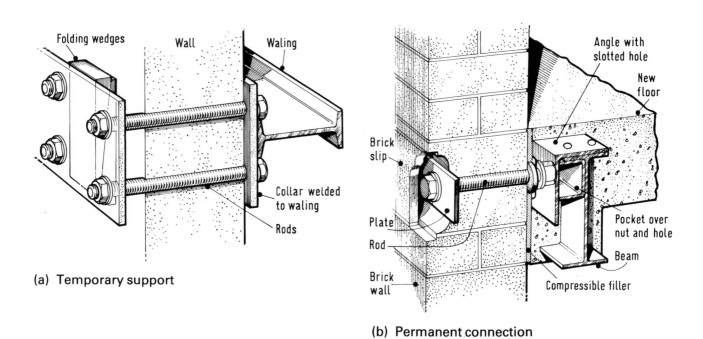

(a) Temporary support

(b) Permanent connection

Figure 63 *Possible means of attaching wall to temporary and permanent supports*

8. Sources of information

Most County, Town or Borough Reference Libraries now maintain a specific local history collection for the area they cover. These collections (or the main reference libraries) contain some or all the following types of information which may be useful in establishing the history of a particular building or site:

geological maps and memoires

historical Ordnance Survey maps

other historical maps - including parish maps

tithe maps, deed plans, etc.

directories

rate books

census returns

local estates' archives

photographs

reminiscences and miscellaneous pamphlets.

8.1   MAIN SOURCES

Brick Development Association

Woodside House
Winkfield
Windsor
Berkshire   SL4 2DX

Tel: Winkfield Row
(0344) 885651

British Geological Survey

Exhibition Road
London   SW7 2DE

Tel: 071-584 4090

Greater London Record Office and
History Library

40 Northampton Road
London   EC1R 0JU

Tel: 071-633 7132

Public Record Office

Ruskin Avenue
Kew
Surrey   TW9 4DU

Tel: 081-876 3444 x2350

Science Reference Library of the
British Library

25 Southampton Buildings
Chancery Lane
London   WC2A 1AX

Tel: 071-323 7494

Society for the Protection of
Ancient Buildings

37 Spital Square
London   E1 6DY

Tel: 071-377 1644

Timber Research and Development
Association

Stocking Lane
Hughenden Valley
High Wycombe
Bucks   HP14 4ND

Tel: (0494) 563091

Other sources may be obtained from the CIRIA Guide to Sources of Construction Information [39].

8.2 OTHER SOURCES

Local Authorities: Planning Department
Building Control Department
Public Health Department (drainage plans).

Statutory Services Undertaker: Local Electricity Board
CEGB
British Telecom
British Gas
Water and Sewerage Authority

References

1. INSTITUTION OF STRUCTURAL ENGINEERS
   Appraisal of existing structures
   1980 (new edition in preparation)

2. KNOWLES, C.C. and PITT, P.H.
   The history of building regulations in London: 1189-1972
   Architectural Press (London), 1972

3. BRITISH STANDARDS INSTITUTION
   Structural recommendations for loadbearing walls
   CP 111:1970 (first version 1948)

4. INSTITUTION OF STRUCTURAL ENGINEERS
   Symposium on structural repair of tenements
   (Scottish Branch), May 1984

5. BRITISH STANDARDS INSTITUTION
   Code of practice for site investigations
   BS 5930:1981

6. LONDON COUNTY COUNCIL
   London County Council (General Powers) Act 1909
   c CXXX

7. LONDON COUNTY COUNCIL
   London Building Act 1930, c CLVIII, 20 and 21 Geo-5

8. BRITISH STANDARDS INSTITUTION
   Code of practice for dead and imposed loads
   BS 6399: Part 1:1984

9. BIDWELL, T.G.
   The conservation of brick buildings
   Brick Development Association, 1977

10. BRUNSKILL, R. and CLIFTON-TAYLOR, A.
    English brickwork
    Ward-Lock (London), 1977

11. FRANCIS, A.
    The cement industry 1796-1914: A history
    David and Charles (Newton Abbot), 1977

12. CLIFFON-TAYLOR, A. and IRESON, A.S.
    English stone building
    Gollancz (London), 1983

13. HEYMAN, J.
    The masonry arch
    Ellis Horwood (Chichester), 1982

14. BUILDING RESEARCH ESTABLISHMENT
    Assessment of damage in low-rise buildings
    BRE Digest 251, 1981

15. BRITISH STANDARDS INSTITUTION
    Structural use of timber
    BS 5268: Part 5:1989
    Code of practice for preservative treatments
    of structural timber

16. BUILDING RESEARCH ESTABLISHMENT
    Dry rot: its recognition and control
    BRE Digest 299, 1985

17. BUILDING RESEARCH ESTABLISHMENT
    Decay in buildings – recognition, prevention and cure
    BRE (Princes Risborough Lab) Technical Note 44, 1969

18. BRITISH STANDARDS INSTITUTION
    The structural use of timber
    CP 112: Part 2:1971 (current version BS 5268: Part 2: 1991)

19. DAWES, M.H. and YEOMANS, D.T.
    The timber trussed girder
    Structural Engineer May 1985 63A(5), 147 to 154

20. BRITISH STANDARDS INSTITUTION
    Fire tests on building materials and structures
    BS 476: Part 20: 1987 Methods for the determination of
    the fire resistance of elements of construction
    (general principles)
    Part 21: 1987 Methods for the determination of the fire
    resistance of non-loadbearing elements of construction

21. BRITISH STANDARDS INSTITUTION
    Structural use of timber
    BS 5268: Part 4:
    Fire resistance of timber structures
    Section 4.1: 1978 Recommendations for calculating fire
    resistance of timber members
    Section 4.2: 1989 Recommendations for calculating fire
    resistance of timber stud walls and joisted floor constructions

22. HISTORICAL STRUCTURAL STEELWORK HAND-
    BOOK
    British Constructional Steelwork Association Ltd., 1984

23. ANGUS, H.T.
    Cast iron – physical and engineering properties
    Butterworths (London), Second Edition, 1976

24. TWELVETREES, W.N.
    Structural iron and steel
    Fourdrinier, London, 1900
    (available for reference at Institution of Structural Engineers)

25. BRITISH STANDARDS INSTITUTION
    Structural steel sections
    BS 4:1903 (current version, 1980)

26. BRITISH STANDARDS INSTITUTION
    The use of structural steel in building
    BS 449:1932 (current version, Part 2:1969)

27. BRITISH STANDARDS INSTITUTION
    Standard specification for structural steel for bridges and
    general building construction
    BS 15:1906, current version BS 4360: 1990 and
    BS EN 10 025: 1990

28. BRITISH STANDARDS INSTITUTION
    Methods for tensile testing of metals
    BS 18: Part 2:1971 (superseded by BS EN 10 002-1: 1992)
    Steel

29. BARNFIELD, J.R. and PORTER, A.M.
    Historic buildings and fire: fire performance of cast iron
    structural elements
    Structural Engineer December 1984 **62A**(12), 373 to 380

30. Fire protection for structural steel in buildings
    Association of Structural Fire Protection Contractors and
    Manufacturers, Second Edition, revised 1992

31. WELTMAN, A.J. and HEAD, J.M.
    Site investigation manual
    CIRIA Special Publication 25, 1983

32. BRITISH STANDARDS INSTITUTION
    Methods of test for soils for civil engineering purposes
    BS 1377:Parts 1-9: 1990

33. TOMLINSON, M.J.
    Foundation design and construction
    Pitman (London), Fifth Edition, 1986

34. BRITISH STANDARDS INSTITUTION
    Code of Practice for Foundations
    BS 8004: 1986

35. WYNNE, C.P.
    A review of bearing pile types
    PSA/CIRIA Piling Group Report PG1, 2nd edition, 1988

36. Town and Country Planning Act 1971 c 78

37. BRITISH STANDARDS INSTITUTION
    Code of practice for falsework
    BS 5975:1982

38. BRITISH STANDARDS INSTITUTION
    Loading
    CP 3: Chapter V: Part 2:1972
    Wind loads

39. CIRIA Guide to Sources of Construction Information
    4th Edition
    CIRIA Special Publication 30, 1984

Bibliography

ASHURST, J. and ASHURST, N.
Practical building conservation
(English Heritage Technical Handbook), Volumes 1 to 5
Gower Technical Press (Aldershot), 1988

ASHURST, J. and DIMER, F.G.
Stone in building – its use and potential today
Stone Federation, Swindon Press, 1984

BAIRD, E.C. and OZELTON, J.A.
Timber designers manual
Granada (London), 1984

BENSON, J., EVANS, B., COLOMB, P. and JONES, G.
The housing rehabilitation handbook
Architectural Press (London), 1980

BOWEN, R.
Grouting in engineering practice
Applied Science Publishers (London), 1981

BRACEGIRDLE, B.
The archaeology of the industrial revolution
Heinemann (London), 1973

BRAY, R.N. and TATHAM, P.F.B.
Old waterfront walls – Management, maintenance and rehabilitation
CIRIA and E & F N Spon (London), 1992

BRERETON, C.
The repair of historic buildings: advice on principles and methods
English Heritage (London), 1991

BUILDING RESEARCH ESTABLISHMENT
Concrete in sulphate-bearing soils and ground waters
BRE Digest 250, 1981

BUILDING RESEARCH ESTABLISHMENT
Cleaning external surfaces of buildings
BRE Digest 280, 1983

BUILDING RESEARCH ESTABLISHMENT
Improving the sound insulation of separating walls and floors
BRE Digest 293, 1985

BUILDING RESEARCH ESTABLISHMENT
Increasing the fire resistance of timber floors
BRE Digest 208, 1988

BUILDING RESEARCH ESTABLISHMENT
Providing temporary support during work on openings in external walls
BRE Good Building Guide 15, 1992

BUILDING RESEARCH ESTABLISHMENT
Repairing brick and block masonry
BRE Digest 359, 1991

BUILDING RESEARCH ESTABLISHMENT
Repairing or replacing lintels
BRE Good Building Guide 1, 1992

BUILDING RESEARCH ESTABLISHMENT
Structural appraisal of existing buildings for change of use
BRE Digest 366, 1991

BUILDING RESEARCH ESTABLISHMENT
Suspended timber ground floors: repairing rotted joists
BRE Defect Action Sheet 74 (design), 1986

BUILDING RESEARCH ESTABLISHMENT
Temporary support for openings in external walls: assessing load
BRE Good Building Guide 10, 1992

BUILDING RESEARCH ESTABLISHMENT
The durability classification of timber
BRE (Princes Risborough Lab) Technical Note 100, 1981

BUNGEY, J.H.
Testing concrete in structures – A guide to equipment for testing concrete in structures
CIRIA Technical Note 143, 1993

CHARLES, F.W.B. with CHARLES, M.
Conservation of timber buildings
Hutchinson (London), 1984

CLIFTON-TAYLOR, A.
The pattern of English building
Faber & Faber (London), Fourth Edition, 1987

CONSTRUCTION INDUSTRY RESEARCH AND
INFORMATION ASSOCIATION
Sound insulation of timber floors in rehabilitated Scottish tenements
CIRIA Special Publications 36, 1987

CRUICKSHANK, D. and WYLD, P.
London: The art of Georgian building
Architectural Press (London), 1975

CURTIN, W.G., SHAW, G., BECK, J.K. and BRAY, W.A.
Structural masonry designers manual
Granada (London), 1982

DAVEY, A.
The care and conservation of Georgian houses: a maintenance manual for Edinburgh New Town
Butterworth Architecture (Oxford), 1988

ELEY, P. and WORTHINGTON, J.
Industrial rehabilitation
Architectural Press (London), 1984

FAIRBAIRN, Sir William
On the application of cast and wrought iron for building purposes
Longman (London), Fourth Edition, 1870

FLEMING, W.G.K., WELTMAN, A.J., RANDOLPH, M.F. and
ELSON, W.K.
Piling engineering
Blackie (Glasgow), 1992

GALE, W.K.V.
Iron and steel
Ironbridge Gorge Museum Trust, 1983

GOODCHILD, S.L. and KAMINSKI, M.P.
Retention of major façades
Structural Engineer, April 1989, **67** (8), 131 to 138

HAMILTON, S.B.
The use of cast iron in building
Trans. Newcomen Society, 1941, **21**, 139 to 155

HARPER, R.N.
Victorian Building Regulations
Mansell Information Publishing Ltd. (London), 1985

HARRIS, R.
Discovering timber framed buildings
Shire Publications (Aylesbury), 1978

HEAD, J.M. and JARDINE, F.M.
Ground-borne vibrations arising from piling
CIRIA Technical Note 142, 1992

HEALTH & SAFETY EXECUTIVE
Evaluation and inspection of buildings and structures
HMSO (London), 1990

HEALTH & SAFETY EXECUTIVE
Façade retention
Guidance Note GS 51
HMSO (London), 1992

HIGHFIELD, D.
Technical problems and their solutions arising from the conversion of large historic buildings
Institute of Advanced Architectural Studies, University of York, 1979

HIGHFIELD, D.
The construction of new buildings behind historic façades
E & F N Spon (London), 1991

HOLLAND, R. *et al.*
Appraisal and repair of building structures: introductory guide
Thomas Telford (London), 1992

HUME, I.
Floor loadings and historic buildings
English Heritage Conservation Bulletin, 1992, 18, 1 to 2

INNOCENT, C.F.
The development of English building construction
David and Charles (Newton Abbot), 1971 (reprint)

INSALL, D.W.
The care of old buildings today
Architectural Press (London), 1973

IRVINE, D.J., and SMITH, R.J.H.
Trenching practice
CIRIA Report 97, 1983

JAGGARD, W.R. and DRURY, F.E.
Architectural building construction, Volumes 1 to 3
Cambridge University Press (Cambridge), various editions
including 1923

JOHNSTON, S.
Bonding timbers in old brickwork
Structural Survey, 1992, **10** (4), 355 to 362

LEEMING, M.B.
Standard tests for repair materials and coatings for concrete.
Part 2: Permeability tests
CIRIA Technical Note 140, 1993

Standard tests for repair materials and coatings for concrete.
Part 3: Stability, substrate compatibility and shrinkage tests
CIRIA Technical Note 141, 1993

LIFF, J.F. and CLAYTON, C.R.I.
Roles and responsibility in site investigation
CIRIA Special Publication 73, 1991

Movement and cracking in long masonry walls
CIRIA Special Publication 44, 1986

LIFF, J.F. and CLAYTON, C.R.I.
Recommendations for the procurement of ground investigation
CIRIA Special Publication 45, 1986

Sound insulation of timber floors in rehabilitated Scottish
Tenements
CIRIA Special Publication 46, 1987

Wall technology (7 volumes)
CIRIA Special Publication 87, 1992

LLOYD, N.
A history of the English house
Architectural Press (London), 1975 (reprint)

MACGREGOR, J.E.M.
Strengthening timber floors
Technical Pamphlet 2
Society for the Protection of Ancient Buildings, 1973

MCGREGOR, J.E.M.
Outward leaning walls
Society for the Protection of Ancient Buildings
Technical Pamphlet 1, 1971

McLEISH, A.
Standard tests for repair materials and coatings for concrete.
Part 1: Pull-off tests
CIRIA Technical Note 139, 1993

MELVILLE, I. and GORDON, I.
The repair and maintenance of houses
Estates Gazettes Ltd., 1973

MICHELL, E.
Emergency repairs for historic buildings
English Heritage/Butterworth (Oxford), 1988

MILLS, R.L., SPYER, G. *et al.*
Aspects of alterations and rehabilitation
Structural Engineer February 1978 **56A** (2), 29 to 58

Discussion on above:
Structural Engineer May 1980 **58A** (5), 163 to 172

MITCHELL, G.E.
Advanced building construction
Batsford (London), various editions

MITCHELL, G.E.
Building byelaws illustrated
Batsford (London), 1947

NEWLANDS, J.
The carpenter's assistant: the complete practical course in
carpentry and joinery
Studio Editions, (reprint), 1990

PARNELL, A.C.
Building legislation and historic buildings
Architectural Press for English Heritage (London), 1987

PITT, P.H. and DUFFON, J.
Building in inner London
Architectural Press (London), 1975

PULLAR-STRECKER, P.
Corrosion damaged concrete – assessment and repair
CIRIA Book B1, 1987

RICHARDSON, C.
AJ guide to structural surveys
Architectural Press (London), 1985

RICHARDSON, C.
Distorted walls: survey, assessment, repair
Architect's Journal 13 January 1988, **187** (2), 51 to 56

RUP, A. RESEARCH AND DEVELOPMENT
Flat roofing: design and good practice
CIRIA Book B15, 1993

SCHAFFER, R.J.
The weathering of natural building stones
Building Research Establishment, reprint 1972

Building design legislation (a guide to the Acts of Parliament and
Government Orders and Regulations which affect the design of
buildings in England and Wales)
CIRIA Special Publication 23, 1982
(updates to July 1985)

Cast iron columns and beams
Greater London Council Development and Materials Bulletin 91,
January 1976

Scottish Building Legislation
CIRIA Special Publication 34, 1985

SOWDEN, A.M. (Ed.)
The maintenance of brick and stone masonry structures
E & F N Spon (London), 1990

SUTHERLAND, R.J.M.
The introduction of structural wrought iron
Trans. Newcomen Society, 1964, **36**, 67 to 84

THORBURN, S. and LITTLEJOHN, G.S.
Underpinning and retention
Blackie Academic & Professional (London), 1993

TIMBER RESEARCH AND DEVELOPMENT ASSOCIATION
Roof space conversions
TRADA (High Wycombe), 1991

TUTT, J.N.
Replacement ties in cavity walls: a guide to tie spacing and
selection
CIRIA Report 117, 1988

Sound control for homes
CIRIA Report 127, 1993

WALLACE, J. and WHITEHEAD, C.
Graffiti removal and control
CIRIA Special Publication 71, 1989

WARLAND, E.G.
Modern practical masonry
Pitman, Second Edition, 1953 (reprinted by the Stone Federation,
1993)

WYNNE, C.P.
A review of bearing pile types
CIRIA Report PGI (2nd Edition) 1988

Selection and use of fixings in concrete and masonry
CIRIA Technical Note 137, 1991

YEOMANS, D.
The trussed roof: its history and development
Scolar Press (Aldershot), 1992